CSAC MONOGRAPHS 6

Nuaulu Ethnozoology
A Systematic Inventory

Roy Ellen

Centre for Social Anthropology and Computing
University of Kent at Canterbury
1993
in co-operation with the Centre of South-East Asian Studies

CSAC MONOGRAPHS

CSAC Monographs is a series produced by the Centre for Social Anthropology and Computing, University of Kent, which aims to make available specialist material in social anthropology and sociology at the lowest possible cost. Production and distribution costs will be kept to a minimum by the use of up-to-date information technology and direct mailing.

The intention is to expand the series as rapidly as possible, and we are willing to consider a wide variety of material including: unpublished dissertations which are still of current interest; edited collections based on conferences and symposia; new editions of important work currently unavailable; and monographs which because of their unusual length or specialist nature are not considered commercially viable by conventional publishers.

Orders and requests for further information should be sent to CSAC Monographs, CSAC Office, Eliot College, University of Kent, Canterbury, United Kingdom CT2 7NS. Editorial enquiries should be sent to the Series Editor, Dr. J. S. Eades, at the same address.

Copyright 1993 by the Centre for Social Anthropology and Computing.

Published and Distributed in 1993 by the Centre for Social Anthropology and Computing, University of Kent, Canterbury, United Kingdom CT2 7NY.

ISBN 0 904938 20 4

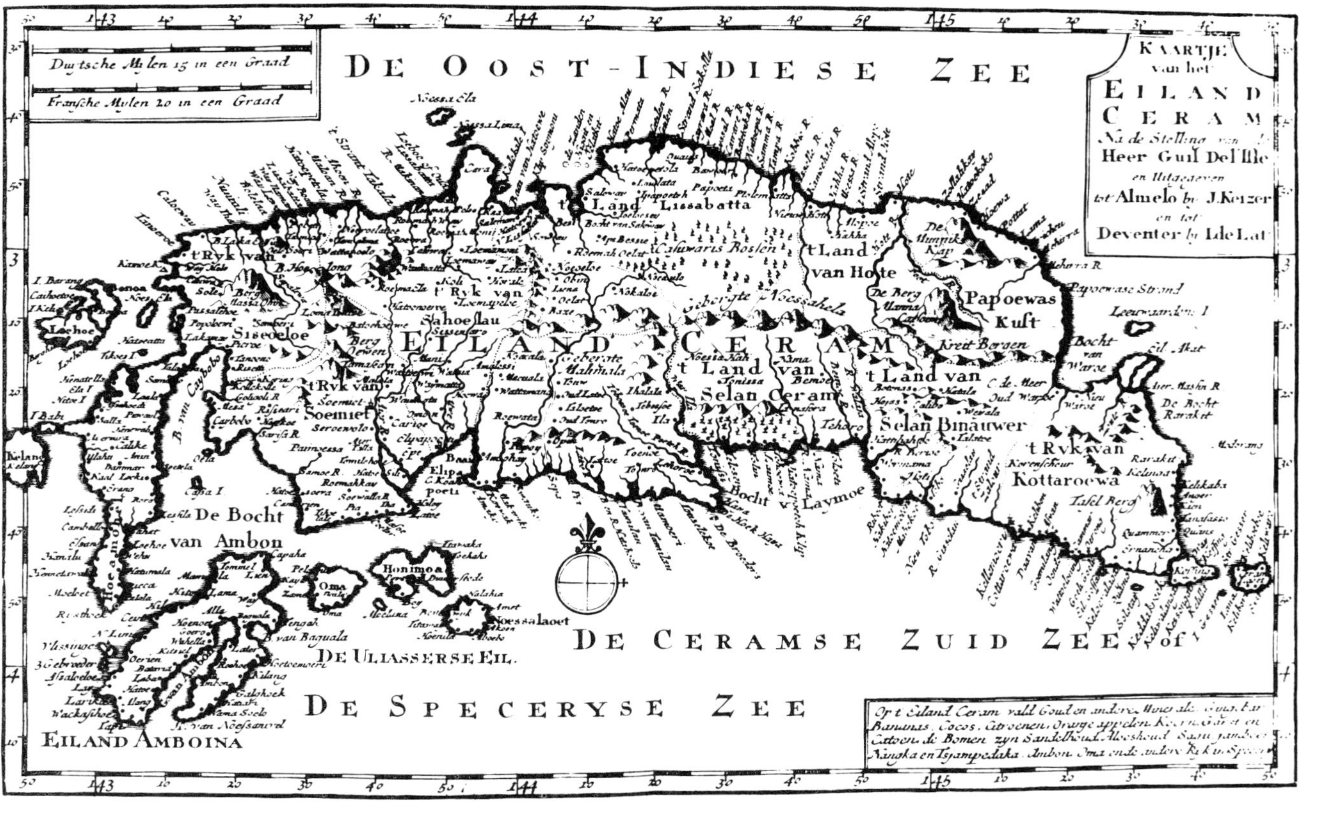

FRONTISPIECE: Map of the island of Seram (Ceram) from the Atlas van Oost-Indien (Almelo; Deventer) published sometime between 1735 and 1747 by Jacob Keiser and Jan de Lat. It is reproduced here by courtesy of the Koninklijk Instituut voor Taal-, Land-en Volkenkunde in Leiden. Note the location of 'Noeoeloe' and 'Noaoala' in the mountains along the north-south path which approximately follows longitude 144E.

CONTENTS

LIST OF TABLES

LIST OF FIGURES

LIST OF PLATES

PREFACE

What follows consists of systematic and annotated lists of categories applied to fauna by the Nuaulu people of eastern Indonesia, the phylogenetic glosses for these categories, plus notes on the uses and cultural associations of the animals concerned. In other words, it is a kind of ethnographic *inventorum natura*. It is designed to complement *The cultural relations of classification: an analysis of Nuaulu animal categories from central Seram*, (published by Cambridge University Press), in which the main arguments, findings and implications of the work undertaken are discussed. The detached presentation of the documents here has been dictated only by their extrinsic character, and by the exigencies of academic publishing. There is also the added advantage of enabling ready reference for those who may wish to use the material for comparison or re-analysis. Each of the chapters group animals in a pragmatic way: in some they are grouped in a way which Nuaulu would recognise as plausible (bats, birds, turtles, snakes, frogs, fish and molluscs), in others the grouping follows Western and scientific convention (lizards, insects, crustacea), in others still the criteria are mixed (land mammals). Throughout this monograph, the guiding principle has been to provide an arrangement and sequence which will be most useful to those who are most likely to want to consult it.

Some of the material included here has previously been published elsewhere, though it generally appears in a revised and modified form. For permission to reproduce copyrighted text, I would like to thank my co-authors, and the editors and publishers of the journals in which the following articles were printed. Chapters 4 and 5 first appeared as 'The content of categories and experience: the case for some Nuaulu reptiles' (with A. F. Stimson and J. Menzies) in *Journal d'Agriculture Tropicale et Botanique Appliquée* (1976, 24: 3-22); and Chapter 7 as 'Structure and inconsistency in Nuaulu categories for amphibians' (with A. F. Stimson and J. Menzies), also in *Journal d'Agriculture Tropicale et Botanique Appliquée* (1976, 23: 125-38). Parts of this Preface and the Introduction also appear in *The cultural relations of classification*, and I am grateful to the Syndics of Cambridge University Press for allowing me to incorporate the text here.

The various periods of fieldwork upon which the study is based have been conducted under the auspices of the Lembaga Ilmu Pengetahuan Indonesia (the Indonesian Academy of Sciences) in Jakarta, the staff of which have always been most generous and cooperative. The 1975 season was also conducted in cooperation with the Lembaga Biologi Nasional (The National Institute of Biology) and the Museum Zoologicum Bogoriense in Bogor, and I am particularly grateful for the sponsorship of Dr. S. Kadarsan in this respect.

Financial support during 1969-71 came from a Social Science Research Council Studentship (No. S68.8243), augmented by grants from the London-Cornell Scheme for research in South and Southeast Asia and the Central Research Fund of the University of London. The 1973 phase was supported mainly by a Hayter Travel Grant. In 1975 I received a Social Science Research Council Award (HR3410.2) for research on 'Nuaulu ethnobiology and ecology', covering the period 1975-77. In both 1969-71 and 1973 audio-visual equipment was provided by the Central Research Fund of the University of London. Acknowledgements are also due to the British Academy, Nuffield Foundation and University of Kent at Canterbury whose assistance on a different project in the same area permitted three further brief visits to south Seram, in January 1981, June 1986 and February-March 1990.

Over the years my ethnozoological work has relied upon the generous help of a large number of specialists. Such support is still unusual in ethnographic research, although very necessary and relatively frequent in ethnobiology. In particular, I have been able to benefit from the expertise of the following staff of the Natural History Museum in London:

Department of Zoology: Mr. John Edwards Hill and Mr. P. D. Jenkins (Mammalia); Miss A. Grandison and Mr. A. F. Stimson (Amphibia, Reptilia); Dr. Alwyne Wheeler, Dr. P. J. P. Whitehead and Mr. O. A. Crimmen (Fish); Mr. K. H. Hyatt, Mr. F. R. Wanless, Mr. Paul D. Hillyard (Arachnida and Myriapoda); F. G. Easton (Annelida); Mr. R. W. Ingle and Dr. Anthony A. Fincham (Crustacea); Ms. K. M. Way, Mr. F. Naggs and Mr. J. F. Peake (Mollusca); and Dr. E. N. Arnold.

Department of Entomology: David R. Ragge, Mrs. Judith A. Marshall (Orthoptera), W. R. Dalling (Heteroptera), R. T. Thompson (Coleoptera), Kenneth G. V. Smith (Diptera), D. Morgan (Hymenoptera), Dr. W. J. Knight (Hemiptera), Ms Julie Harvey (Library), and Ms Theresa Clay.

Sub-Department of Ornithology (Tring): Dr. D. W. Snow, Dr. P. J. K. Burton and Mr. G. Galbraith.

Other specimens were identified for me by the late Dr. Serene in Paris, Mr. J. Menzies and the staff of the Department of Biology at the University of Papua New Guinea in Port Moresby, and Dr. Soenartono Adisoemato in Bogor. I am also grateful to the British Museum (Natural History) and the

Biology Department of the University of Kent at Canterbury for the provision of collecting equipment and preservatives. Dr. J. D. Kesby of the University of Kent and the late Dr. C. M. N. White of Lytham St. Annes were most helpful in providing informed comment on a number of queries. While this monograph is not an account of Nuaulu ethnobotany, the identification of plant species has been a necessary part of a broader understanding of Nuaulu relations with their fauna. I am consequently also indebted to Mr. L. L. Forman of the Royal Botanic Gardens at Kew, and Dr. Chang Kiaw Lan of the Botanic Gardens in Singapore.

Finally, I would like to thank all those who have supported me during various fieldwork phases: in Amboina, Sepa and the Nuaulu villages of Hahuwalan, Watane, Aihisuru, Bunara and Rohua. Since 1988 I have been able to draw on the work of Rosemary Bolton of the Summer Institute of Linguistics; it is her orthography which I adopt here in almost every instance, while her clarification of various details of nomenclature, semantics and grammar have been invaluable. Jane Pugh kindly modified some existing maps, Brian Durrans of the Museum of Mankind in London has facilitated access to Nuaulu artifacts, while G. A. Nagelkerke, Mrs. L. van der Spree-Annyas and Gerrit Knaap have courteously handled a number of queries. For expert editing assistance in producing camera-ready copy I am indebted to Barbara Delaney, Michael Fischer and Jan Horn. Various chapters have benefitted from the critical comments of the late Ralph Bulmer and Paul Taylor, though neither would necessarily approve of the final product in its entirety.

R. F. Ellen,
Crockshard Farmhouse, Wingham

A NOTE ON ORTHOGRAPHY

As so much of my argument here hinges on the form and meaning of Nuaulu words, orthography is no insignificant matter. My own language materials are extensive, but unsystematic and linguistically unsophisticated. It is therefore a great pleasure, and something of a relief, to be able to draw upon the recent work of Rosemary Bolton [1990], work which at the time of writing is still in progress. The letters which she uses to represent Nuaulu speech sounds are phonemic, and include 11 consonants and five vowels composed of the following phonetic features:

Consonants

Glottal				
Voiceless stops	p	t	k	
Fricatives		s		h
Nasals	m	n		
Flaps		r		
Laterals		l		
Semi-vowels	w	y		

Vowels

		Front Unrounded	Back Rounded
High	close	i	u
Mid	half open	e	o
	open	a	

The Indonesian alphabet includes all Nuaulu phonemes and is used here without modification. Stress is unmarked in most regular cases and normally occurs on the penultimate syllable. Indigenous words appear in **bold-face**. One consequence of my adopting this revised phonology and orthography, as alert readers may notice, are certain changes in the written appearance of some Nuaulu words (e.g. **totuwe** becomes **totue**). There is not as yet any Nuaulu consensus as to the proper way to write personal names, clans and places. To avoid confusion they usually appear here as in earlier publications by me.

CHAPTER ONE

INTRODUCTION

This monograph provides much of the data upon which the arguments contained in *The Cultural Relations of Classification* rest. As I have explained in the Preface, it cannot be considered independent of this and for reasons of economy much background data are omitted here: data on the Nuaulu language, particularly on the structure of animal nomenclature, and on various procedural matters. Those who wish to make fuller sense of what is here presented are strongly advised to consult that work. On the other hand, some basic ethnographic orientation is essential, and this accounts - for example - for the reproduction of maps which appear in the companion volume. The main purpose of this introduction is to provide this orientation, to say something about the history of ethnozoological studies on Seram, my own fieldwork and the methods employed, and to explain the technical conventions adopted in the chapters which follow.

1.1 The Nuaulu: culture, society and environmental context

The island of Seram lies in the central Moluccas (figure 1), part of the modern Indonesian province of Maluku. It is, therefore, part of the biogeographic region of Wallacea, a term used here to conveniently designate those islands lying between the Sunda and Sahul continental shelves [Darlington, 1957: 462-73; Ellen, 1978b; White, 1973: 175]. In biogeographic terms, it marks a zone of transition between the oriental biota of southeast Asia and that of Melanesia, Australia and beyond. Its western boundary is marked by Wallace's faunal line, its eastern boundary by Lydekker's line. Because it is a transitional zone of small islands the fauna is, for many land-based groups, a relatively depauperate one.

The larger islands of the Moluccas (Seram, Halmahera, Buru) are still dominated by tropical rain forest, although most of the smaller ones are now denuded and extensively planted with clove, nutmeg, coconut palms and other useful trees, and subjected to forms of dryfield cultivation, especially along level coastal land.

Seram itself (figure 2, plate 1) can be usefully divided into about ten terrestrial biotopes [see also Ellen, 1984: 177-9]: (1) montane forest, (2) mature

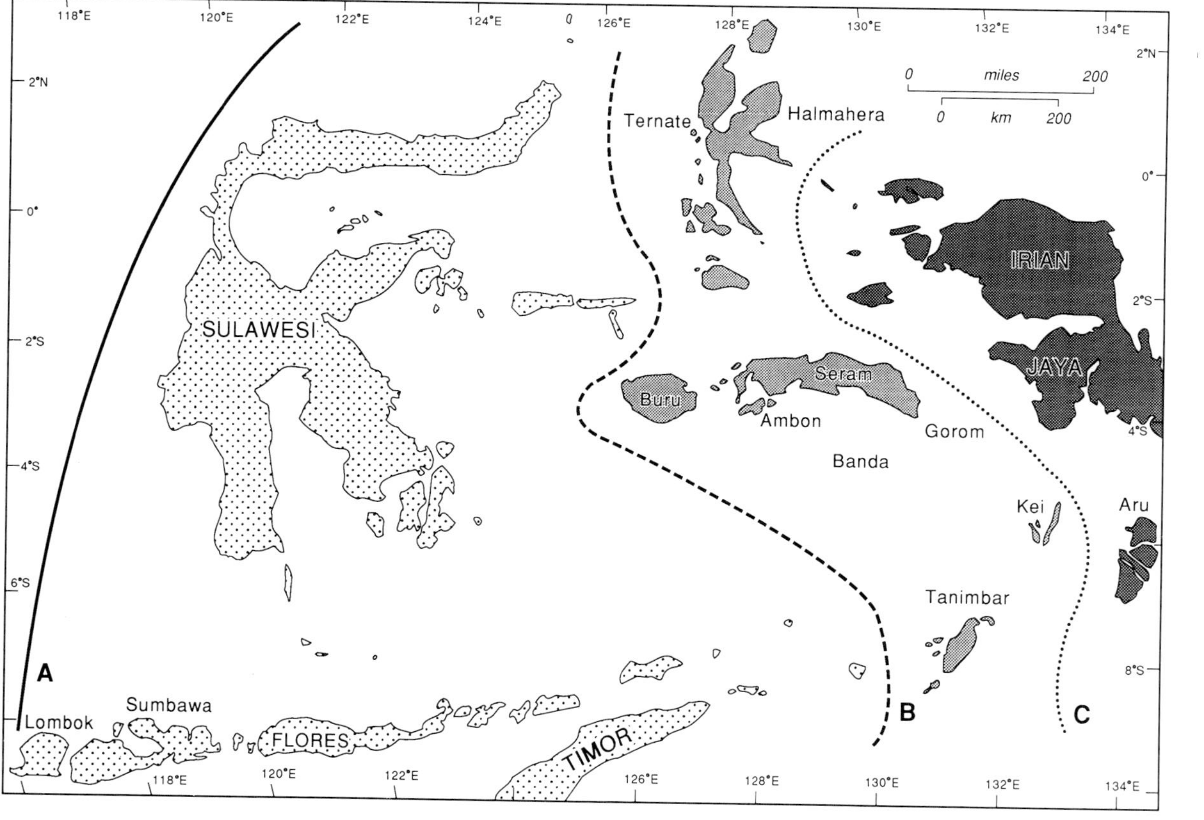

FIGURE 1: The Moluccan islands in relation to Sulawesi and Irian Jaya, showing (a) Wallace's line of faunal balance, (b) Weber's line and (c) the western boundary of the Australian biogeographic region. Wallacea is the aera between lines *a* and *c*. The modern Indonesian province of Maluku includes all islands between lines *b* and *c*, plus the Aru archipelago to the southwest, Sula to the west and Wetar to the southwest. Note: Unless otherwise stated, the orientation of all maps is identical to fiure 1

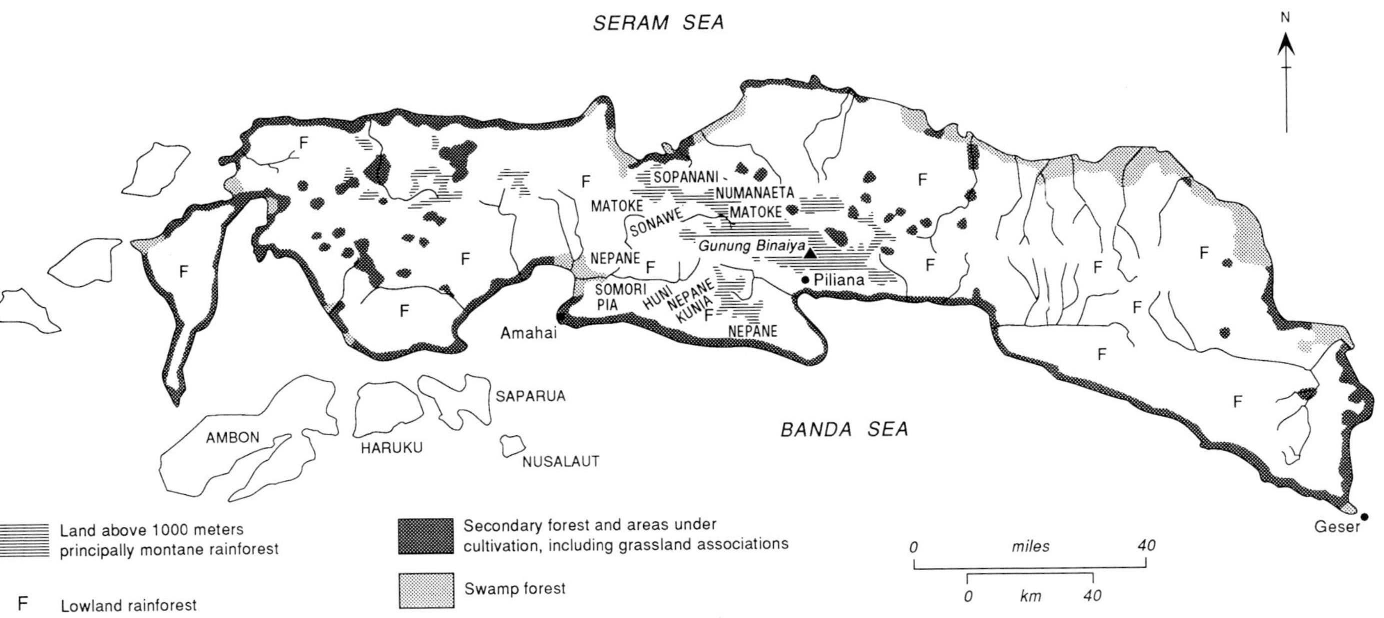

FIGURE 2: The island of Seram and adjacent areas, showing population distribution and principal environmental zones. Elpaputih Bay lies W of Amahai, Teluti Bay SE of Piliana and Seleman Bay half way along the north coast. Adopted from Topographische Dienst 1921, *Eiland Seran en omliggende eilanden* (Scale 1: 5000 000), The Netherlands. The approximate location of Nuaulu clans prior to their movement to the coast, and as elicited from informants, has been superimposed (e.g. MATOKE). Square brackets indicate the bottom corners of the area enlarged in figure 3.

lowland rain forest, (3) mixed secondary forest, (4) bamboo brush, (5) sago swamp forest, (6) swidden and dryfield cultivated areas, (7) planted grove-land, (8) freshwater rivers, streams and pools, (9) grassland, and (10) settlement sites. To this we can add a further four coastal and marine biotopes: (11) the littoral, (12) rocky shores, (13) sandy shores, (14) muddy shores, mangrove swamps and estuaries, (15) coral reefs, banks and atolls, (16) deep sea waters. These variations in the environment with which Nuaulu interact are illustrated in plates 2 - 5 here, and also in plates 1.1 - 1.6 of *The Cultural Relations*.

The Nuaulu are a people of the south central part of the island. They inhabited, during the period of my research, five settlements along the narrow coastal strip, in the vicinity of the old Muslim domain of Sepa (figures 2 - 3). This is approximately where longitude 129 5' East meets the south coast of the island, in the Amahai kecamatan (administrative sub-district), between the bays of Elpaputih and Teluti. In 1971 they numbered some 500 individuals, and formed approximately half of the speakers of the 'Nuaulu' language group; the remainder living in and around the villages of Oping and Rumaolat on the north coast at Seleman Bay. The population has almost doubled in the two decades since then.

Before the middle of the nineteenth century Nuaulu clans occupied separate hamlets in the highlands around the drainage systems of the rivers Ruatan and Nua (figure 3). At this time there had been relations between coastal Muslim settlements for at least two hundred years, and the Nuaulu were clearly engaged in intermittent relations of enmity (head-hunting and warfare), alliance, and probably trade [1]. In this respect the situation must have been similar to that of other highland peoples on Seram at that time. Towards the end of the nineteenth century the Nuaulu began to occupy sites near to Sepa, apparently under pressure from the Dutch and the coastal rajas. This movement is significant because it provides us with a shift in environmental and economic conditions which might be hypothesized to have had radical implications for their knowledge, classification and use of local fauna. At the present time, Nuaulu ecological and economic relations remain, significantly, oriented to the mountainous interior of the island, rather than to the coast. The most important starch staple is *Metroxylon* sago. This is largely extracted from forest palms, though is sometimes planted. The forest is also the source of most animal protein, much vegetable food and materials for manufacturing and other technical purposes. Swiddens are cut each year within a four kilometer radius of the village, are energy intensive, but contribute a disproportionately small amount to the total diet [Ellen, 1988a]. Garden crops are varied, but with manioc, taro, sweet potatoes, yams, bananas and plantains predominating. Prior to the

PLATE 1: NASA Landsat image of Seram for the months September-November. The image for the western half of the island is dated 4 October 1972, that for the eastern half 24 January 1979. Scale 1:1 000. I would like to thank Mr J R Marshall of Clyde Surveys for his help in obtaing this print.

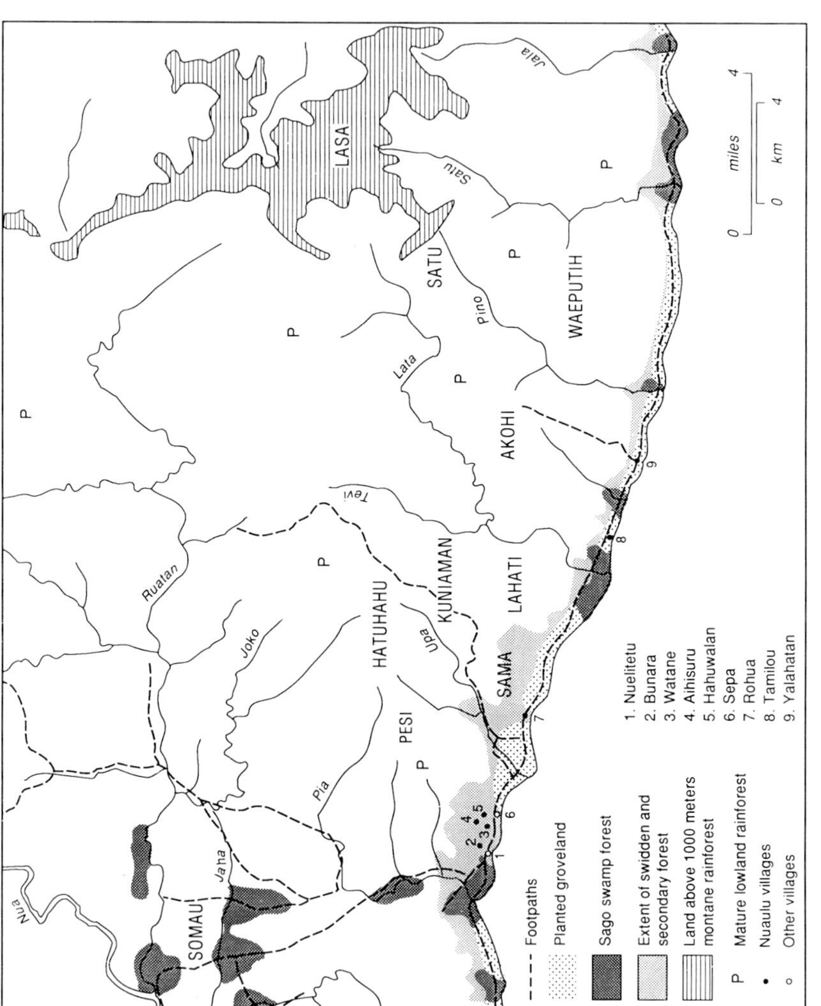

FIGURE 3: Location of villages and approximate distribution of major terrestrial biotopes in the Nuaulu extractive area, as of 1971. Names in upper case lettering indicate specific areas of forest distinguished and used by the Rohua Nuaulu. It should be noted that commercial lumbering activities and government resettlement schemes from 1975 onwards have now considerably modified the pattern of land use. Adopted from Topographische Inrichting 1919, *Schetskaart van Ceram* (Scale 1:100 000), Batavia; modified on the basis of field cartographic data and notes.

1939-45 war most cash was obtained by the Nuaulu through the collection and trade of forest products, such as dammar resin (*Agathis dammara*), but since Indonesian independence Nuaulu have become increasingly involved in the growing of clove and coconut palms for copra. Animal domestication for food (mainly fowl) is virtually non-existent, though dogs are kept for hunting.

Prior to resettlement, the Nuaulu clan (**ipan**; alt. **ipa, ipane**) was an autonomous patrilineal and exogamous descent group occupying a single hamlet. The **ipan** still retains considerable independence in ritual, political and economic matters, but except for the village of Hahuwalan all villages in 1971 consisted of five or more clans. The Nuaulu first became incorporated within the administrative structure of Sepa, as a separate 'soa', for a period under their own raja, and around 1882 became part of the 'Onderafdeeling' of Amahai. By the time of Indonesian independence there were three Nuaulu administrative units, though five physically-distinct hamlets. Each of the administrative villages of Bunara, Niamonae (in Malay, 'Nuaulu Lama', containing the separate hamlets of Watane, Hahuwalan and Aihisuru) and Rohua[2], were accorded a permanent government head (' kepala pemerintah') in a state-imposed scheme.

Each **ipan** is divided into two **numa**, descent groups focussed on a ritual house and headed by either an **ia onate ipan** or **kapitane**. The **numa** of each clan are in ritual opposition to each other [Ellen, 1986: 7-8]. Marriage is ideally between bilateral cross-cousins, and therefore functionally consistent with relationships between pairs of clans which may endure over many generations. However, any one clan is likely to have relations of marital alliance with many others, while marriage with actual classificatory cross cousins is rare. Clans are theoretically equal in their status, an arrangement which matches an ideology of prescriptive bilateral cousin marriage, traditional clan autonomy and the absence of an overarching indigenous political authority. The clan Matoke, nevertheless, is a ritual *primus inter pares*, providing as its headman what in Ambonese Malay is known as the 'tuan tanah', or Lord of the land.

During the periods of fieldwork on which this monograph is based, research has been centred on the village of Rohua, which in 1971 had a population of 180. The area from which the Rohua Nuaulu extracted during these times - some 900 square kilometers - broadly coincides with the drainage basins of three major watercourse systems: the Ruatan, Pia and Lata-Tevi (figure 3). It is to this which I refer when I speak of 'the Nuaulu area': the greater part of the land mass between Elpaputih Bay in the west and Teluti Bay in the east, south of the watershed formed by the central

PLATE 2: Young primary forest on the river Usa, just prior to clearing: 28 January 1971 (neg. 71-17-25). The dense ground cover is an indication of recent thinning.

PLATE 3: Cultivated land of two years standing near the river Usa, showing secondary forest in background and recent regrowth of **hunua** in mid-ground: 22 August 1973 (neg. 73-5-19).

PLATE 4: Groveland owned by Inane Matoke on the river Awau and planted with mature coconut palms: 9 August 1973 (neg. 73-3-4).

mountainous spine of the island. All of the terrestrial biotopes described earlier are found in this area and their composition is familiar to the Nuaulu. The area includes the heavily populated and cultivated coastal strip, secondary forest surrounding areas of settlement, mature lowland rainforest and swamp forest. Typical lowland rain forest stretches from sea level mountainwards, and dominates the overall ecology of the area. Montane and submontane rain forest, is found above 1000 m., but the only occasions on which this zone is traversed by Nuaulu is during journeys to north Seram, on the longer hunting expeditions to the headwaters of the Nua, Ruatan, Kawa or Lata, or in collecting resin from the conifer *Agathis,* a particularly prominent feature of higher areas on the southern slopes of central Seram.

In composition and structure lowland evergreen forest (mostly on low hill land) is typical of adjacent afforested areas of Southeast Asia and New Guinea, though species diversity is low, 10 - 30 species per hectare not being uncommon. In the south central part of the island the dominants of the more inland and highland areas, representatives of the families Fagaceae, Lauraceae, Icanaceae, Ericaceae (rhododendrons), together with *Agathis,* are replaced towards the coast by *Shorea, Canarium sylvestre, C. indicum,* species of *Terminalia, Calophyllum, Myristica,* and hardwoods such as *Pterocarpus* and *Diospyros.* Up to 50 percent of the volume over 35 cm diameter are Dipterocarps. There is one species of Eucalyptus, *Eucalypt deglucta.* In favourable localities the giant *Melaleuca cajuputi (= leucodendra),* and *Melastoma (Malabathricum?),* are common, as are stands of *Ficus* and *Casuarina* in riverine areas. These same lowland areas contain both permanent and seasonal swamp forest, providing a valuable source of naturally-propagating *Metroxylon*[3].

The human population density of Seram is low - about 0.07 persons per hectare. As a result there has been relatively little succession to anthropogenic grasslands, particularly in the Nuaulu area. Level coastal land is heavily cultivated with a considerable proportion devoted to cash-cropping of cloves and coconuts. The land rapidly steepens away from the shoreline and even coastal populations such as the Nuaulu are forced to cultivate gardens on the steep valley walls of the short rivers descending to the sea. In some localities there are more extensive areas of low-lying land, such as around the mouth of the Ruatan river on Elpaputih Bay. Some of this has been used for garden land and in places has succeeded to grassland. Other areas are too swampy. Of the marine biotopes, the Nuaulu have an intimate knowledge of only the littoral and rocky and sandy shores characteristic of the coastline immediately adjacent to their villages, and no firsthand knowledge whatever of deep sea.

PLATE 5: River Lata between Rohua and Tamilou, at the height of the rainy season: 17 July 1975. Note the tall stems of the *Saccahrum* grass along the banks (neg. 75-1-33).

PLATE 6: Women collecting small fish and molluscs in rock pools as the high tide recedes. Mouth of river Mon, 1 km. east of Rohua: 2 September 1973 (neg. 73-6-21).

1.2 The descriptive ethnozoology of Seram: an historical prolegomenon

The first and most celebrated non-native naturalist of the Moluccas was Georgius Everhardus Rumphius (1627-1702). Rumphius discusses Seramese animals in his two major posthumously published works: *D'Amboinsche Rariteitamer* (which first appeared in 1705), and *Het Amboinsche Kruid-Boek* (which first appeared between 1741 and 1755, and is often known as the *Herbarium Amboinense*). The first of these covered crustacea, echinoderms, corals and molluscs; the second (although ostensibly a treatise on plants) contains many stray notes and observations on animals, particularly insects. In these works he provides some excellent descriptions and illustrations based on observations of species in the field, together with accurate provenances for the animals reported. Although Rumphius completed a further major manuscript on animals (the *Amboinsch Dier(en)boek*), it was subsequently lost, though not before François Valentijn had been able to draw upon it extensively for his *Oud en Nieuw Oost-Indiën*. To say that Valentijn was 'able to draw upon it', may, however, be misleading, since there is substantial evidence that he reproduced sections more-or-less verbatim. Perhaps unfortunately for us he did so with some carelessness, while sometimes altering the wording to suit his biases. The most important section in the *Dier(en)boek* appears to have been that on birds, although there are also descriptions of mammals, fishes, amphibia and reptiles.

More than 150 years separate Rumphius and the next important naturalist of the Moluccas, Alfred Russel Wallace, who travelled and collected in the region during the mid-nineteenth century. His *The Malay Archipelago* [1869] contains a large section on Seram and its offshore islands. Henry Ogg Forbes [Ellen, 1978a; Forbes, 1885], who traced Wallace's footsteps some years later, has a few notes on central Moluccan natural history, but his most important work was conducted on Tanimbar to the southeast. Between 1852 and 1873 the Dutch scientist P. Bleeker published 26 papers on the fish fauna of Ambon [Weber and Beaufort, 1913-1922], and collected some 78 species. From the middle of the nineteenth century to the early years of the twentieth century many books and papers were published on Moluccan avifauna. Wallace, of course, was a notable contributor in this respect, but except for Sieber's treatise on the birds of Buru [1930], and some brief papers by other authors, not a single overview of Moluccan ornithology appeared until the publication of Van Bemmel's list in 1948. Salvadori had published a classic handbook [1880-1882], but this covered a much wider area and he had no firsthand experience of the Moluccas. The Siboga expedition, in 1899-1900, assembled an impressive collection of zoological data on eastern Indonesia, and after 1901 a large number of monographs appeared by authors of different nationalities arising from that expedition. Professor Dr. L. F. de Beaufort edited, and partly wrote, a series of articles dealing with several classes of animal collected during the years 1904-10 on Seram and elsewhere, appearing in 1913. Together with Max Weber (and latterly other collaborators), he wrote the classic *The fishes of the Indo-Australian Archipelago* [11 volumes, 1911-1962], dealing with many fishes of the Moluccas[4].

From 1970 onwards I have spent an increasing proportion of my Nuaulu fieldwork assembling ethnobiological data. All told, this has been spread over three main research periods: 18 months between December 1969 and June 1971, three months in 1973, a further three months in 1975, plus a short two-week visit in January 1981, ten days in June 1986 and three weeks in February-March 1990[5]. The 1975 season was a particularly decisive one since part of my time was spent working with James Menzies, an academic zoologist, who had already cooperated extensively with Ralph Bulmer [e.g. Bulmer and Menzies, 1972-3b; Bulmer et al, 1975]. By June 1971 it was already clear to me that I should concentrate on an examination of Nuaulu knowledge, uses and classification of animals. The decision was made for essentially practical reasons: the corpus had to be limited in some way in order to permit as full and detailed a study as possible, and it seemed to me at that stage that I was in a better position to provide complete and accurate identifications and information on animals than on plants.

The present work lists systematically animals reported for Seram in the Nuaulu area and (for the most part) of which the Nuaulu claim some knowledge. Here it is only necessary to note two general features of the zoology. The first is that, for geographic reasons, the land vertebrates of Seram and the Moluccas in general are surprisingly little differentiated, with the possible exception of birds and murids. This is certainly so when compared with New Guinea. The frogs include no endemic genera and few endemic species. Reptiles, murids, bats and also birds seem to include more distinct endemics. This comparison is in part subjective, being based on our rather limited current knowledge of this area, but there can be little doubt about the low level of endemism for the non-murid terrestrial mammals of the lesser Sundas [Ellen, 1978b: 1.4-7]. Such a low species diversity index has obvious implications for understanding Nuaulu classification of animals, and I shall return to it in a discussion of the relationship of Nuaulu categories to biological species, and in the systematic accounts of the knowledge for each faunal group. The second feature is that in comparing lists for definitely recorded genera and species on Seram as a whole with the equivalent lists for the Nuaulu area in particular [*ibid.*: tables 1 and 2], the number of lizards, snakes, marsupials, bats, even-toed ungulates and rodents recorded for the Nuaulu area is less than the number for Seram as a whole, in some cases strikingly so. It may be that certain species are actually unrepresented in the Nuaulu area, or that provenances given in the older literature and museum collections are doubtful, while older and poorly-known species often turn out to be only subspecies or varieties. For these and other reasons - including the very arbitrariness of the species concept - competent zoologists are justifiably circumspect in the reliance they attach to 'number clues' to evolution and dispersion obtained through quantifying taxa [Darlington, 1957: 31-2]. However, these things are unlikely to explain satisfactorily wide discrepancies, as in the case of bats. In this and other cases it is probably simply that collections have been insufficiently exhaustive. Animals in niches only rarely visited by the Nuaulu are less likely to come to light during the course of fieldwork, and some species present in the area may well be unknown to informants. There is, for example, the Seram island bandicoot (*Rhynochomeles prattorum*) which occurs in the upper limits of the Nuaulu extractive environment, above an altitude of 1000 meters. Although it is allegedly known to, and hunted by, the mountain villagers of the Manusela area, from where it was first recorded. I have no conclusive evidence that it is known to the Nuaulu, despite repeated enquiries during six stretches of fieldwork.

All word lists collected on Seram have, from the earliest times, contained some terms for animals. Likewise, most accounts of fauna have also contained information on local names, uses and beliefs. Rumphius, for example, must be counted as an important source for Ambonese ethnobiology. Not only does he provide us with early Ambonese Malay and other local terms, but he also gives details on animals as pests and as sources of economic products, on veterinary science, on their medical uses, and local beliefs concerning them. But the ethnobiological interest in Rumphius' work lies not only in what he reports concerning the Ambonese, but in the way in which Ambonese and Malay conceptions of knowledge influenced his own ideas, and through them the subsequent development of scientific nomenclature and taxonomy [Peeters, 1979].

For the most part, word-lists collected by naturalists rather than linguists have been the more extensive, more accurately glossed and more ethnobiologically informative. Wallace, 1962 (1869) includes many useful folk biological observations and word lists. Ribbe, who is mostly noted for his 1884 entomological work on Seram [Ribbe, 1889], recognised the crucial adaptive significance of native knowledge of indigenous fauna [Ribbe, 1892: 175-6], as well as continuing the tradition of inter-meshing ethnographic observations with a more general description of his visit. The combination of zoological and linguistic competence is rare enough, but when it occurs the results are generally immensely fruitful. From a folk-zoological point of view, the ornithological work of Edwin Stresemann is important [Stresemann, 1914], almost certainly because of his complementary linguistic interests [Stresemann, 1927]. For the first time we are provided with a large number of local Seramese terms for different kinds of bird, accompanied by specific provenances and careful taxonomic identifications. Among Stresemann's terms are 19 for the Nuaulu.

1.3 Obtaining zoological data

It is no longer excusable for ethnographers to collect local animal terminologies without having taken all reasonable steps to establish their accurate scientific glosses. Without such information, the value of detailed data on their habits, uses, ritual aspects and general folklore is diminished, and may sometimes be quite useless for subsequent workers. Even if we conclude that phylogenetic categories[6] are no more 'real', and as much a part of a folk traditions as those of the Tzeltal, Kalam or Nuaulu, they still provide a baseline for ethno-linguistic description and analysis, an indispensible 'etic grid' [Hunn, 1975: 309] for cross-cultural comparison. The importance of this exercise is not simply to highlight curious contrasts with the 'scientific' view of the universe, or to amaze ourselves with the similarities between folk and phylogenetic models, but because biological taxonomy provides a highly convenient and (ideally) a universally consistent reference plane which can be applied cross-culturally and through which one folk practice can be compared with another. It allows the analyst to avoid the hopeless relativism associated with the view that because our minds are socially-constructed, objective knowledge can be no more than a chimera [Ellen, 1978b: 142]. Furthermore, there is no economical way of glossing folk categories *except* by using scientific zoological identifications [Bulmer, 1969: 4].

Provision of actual specimens adds vital stimulus material when discussing ethnozoological matters. One can no more think of studying animal classification without animals than one might study colour classification with descriptions of colour rather than examples. Thus, questions in the abstract will seldom elicit more than a basic set of widely-shared terms, and will be particularly poor in terminal categories. The presentation of actual specimens invites detailed examination and identification. Moreover, if - as happened to me on many occasions - you repeatedly obtain just one common species of a genus more widely represented, you may equally be given 'generic' terms by an informant. Only when a different species is presented are names for terminal categories revealed for both types. This was the case with Nuaulu **makasisi popole,** a term used for various genera of robust dragon-flies and related forms. Similarly, it is common to initially obtain a list of names for terminal categories which are clearly closely related, but whose relationship only becomes disentangled when actual specimens begin to turn up; while synonyms may be treated as referring to separate categories unless linked to actual specimens. The careful collection, identification and full documentation of animal specimens has, throughout, been the most important and time-consuming aspect of my ethnozoological work.

Specimens were mostly obtained in the vicinity of Rohua, at sea-level. They seldom came from outside the area demarcated by the borders of the most distant gardens, that is from the *inner* area of the Nuaulu extractive environment. However, during late July 1975 small collections were made near the Jala river estuary at sea-level and at nearby Tohai, in the non-Nuaulu village of Piliana at an altitude of 700 m in the central highlands north of Teluti Bay, and at Somau (Tihun), an area of swamp forest towards the mouth of the Ruatan river frequented by the Nuaulu for the purpose of collecting wild sago. Jala, Tohai and Piliana are well-outside the Nuaulu extractive environment (figure 2), and specimens collected by Menzies, with the help of local assistants, were only later discussed with Nuaulu informants in Rohua.

Most of the procedures adopted for the collection of specimens have followed the recommendations laid down by Bulmer [Bulmer and Tyler, 1968: 335-7; Bulmer, 1969], and I have tried to emulate and adapt the rigorous data production techniques used by Berlin et al, 1974 in their study of Tenjapa Tzeltal ethnobotany. I have also added to the repertoire of techniques discussed by Bulmer, but have only modified them when special interests and local conditions have made it imperative that I should do so.

Whenever conditions permitted, and when equipment was available, animal specimens were collected. Most were obtained for me by informants in the course of their normal daily activities. In some cases I was present when an animal was caught, and in such situations was able to make detailed notes on habitat and behaviour. When this was not possible, informants had to be relied upon for information regarding the context of capture, and in some cases of killing. Nuaulu collectors were asked to provide information on the location, behaviour and habitat of animals. I have no reason to believe that this was a particularly unreliable procedure, and in many cases I was able to visit the site of capture afterwards to check up on the circumstances of encounter and other contextual information. The conditions of fieldwork, and perhaps also my own preferences, have resulted in more complete documentation for (as far as vertebrates are concerned) reptiles, amphibians and mammals, than for birds and fish. Consequently, the accompanying ethnozoological data are better for the first three groups than for the rest, something which is inevitably reflected in the weight of illustrations employed.

Most animals were live when brought to me by informants, but a proportion were damaged, and some already dead (either on discovery or by the time they reached me). In 1975 small mammals (primarily murid rats) were obtained by setting baited live and break-back traps overnight, while lamps and mist nets of fine nylon stretched across entrances to caves were used to collect bats and swiftlets. As it happens, the traps did not prove very effective. Other animals, such as frogs, were caught live by hand or with hand-held nets, by both Ellen and Menzies, and subsequently discussed with informants.

Although some small specimens were killed in the field by body compression, most larger specimens were killed by injecting 100 percent Euthatal or MS 222-Sandoz into the heart and then fixed and preserved in a 10 percent solution of formalin through injection and immersion. Wet specimens were usually wrapped in muslin dampened with preservative and stored in polythene bags, small glass phials (in the case of larger insects) or plastic containers in which they were despatched to identifying institutions. Some birds and small mammals were preserved dry as skins by Menzies, and a number of larger specimens were prepared as skeletal material, either as skulls or in their entirety. Before skeletal material was prepared measurements of the full carcass were made. Some crania and mandibles (mainly deer and pig, but also cuscus, cassowary and some fish) kept by the Nuaulu as trophies were obtained where possible (plate 14; also Ellen 1993: plate 1.7), and several hundreds more measured. Most small invertebrates were preserved partially or wholly in 15:1 formalin or (by preference) in 70 percent alcohol. Most Lepidoptera, Coleoptera and other insects were preserved dry, and in the case of molluscs only shells were retained for subsequent identification. A few endo- and ecto-parasites of larger species were collected separately in formalin. The specimens were identified at the Natural History Museum in London, the Biology Department of the University of Papua New Guinea in Port Moresby and at the National Institute of Biology in Bogor, in which institutions they are also deposited.

Photographic records and sound recordings were obtained whenever practicable, and for some groups (e.g. fishes, large mammals and birds) these have been important for subsequent identification and analysis. Many more specimens were examined and then discarded or released than were preserved. In particular, I was not willing to entertain large collections of Seramese avifauna which might involve the gratuitous destruction of an environment which is already threatened by excessive logging and indiscriminate commercial bird-hunting. Though officially protected, the salmon-crested cockatoo, *Cacatua moluccensis,* is actively hunted for sale, as are black-capped or purple-naped lories (*Lorius domicella*). The endemic long-crested myna, *Basilornis corythaix,* and Forsten's oriole, *Oriolus forsteni,* are uncommon to rare, and probably endangered [Amir and Wind, 1978]; while the Moluccan scrubfowl, *Megapodius wallacei,* and the the Nicobar pigeon, *Caloenas nicobarica,* are seemingly also threatened [White and Bruce, 1986: 170].

Table 1 shows the total number of specimens examined from various phylogenetic groups collected in the field. A relatively small number of specimens were retained for later examination and identification outside the fieldwork locality. Most were released live, some were retained in preservative and later discarded because they represented superfluous duplicates of common and easily identifiable species (e.g. *Litoria infrafrenata, Mus musculus, Hemidactylus frenatus*), or because particular species were felt to be endangered. Still others were rejected on the grounds that they were so mutilated or decomposed as to be unsuitable for further use. Also, many specimens were unfortunately lost in transit, or have since gone astray at their institutions of deposit. The figures include duplicates where earlier specimens were considered to be in poor condition, where there was a presumption of significant racial or other taxonomic variation or where Nuaulu informants provided different terms. Small insects such as flies, bugs and beetles were often collected in multiples at a time, and given a single serial number. Consequently, the figures for smaller invertebrates are for 'lots' rather than for individual organisms. Other specimens consisted of parts of larger animals, mainly crania.

1.4 Ethnographic research procedures

Basic guidance on both the biological and ethnographic aspects of ethnozoological research procedures is provided by Bulmer [Bulmer, 1969; Bulmer, 1974]. I have been very much influenced by his sound empiricism, but have tried to augment this with the introduction of some more formal techniques. During 1969-71 the elicitation of data was prolific, but frankly haphazard and opportunistic. In the 1975 season methods employed were more systematic. Informants were first asked to name all animal categories that they knew, irrespective of degree of inclusiveness. Both systematic zoological and cultural data were transferred to edged punch-cards (Copeland-Chatterson Paramount form CC1, 102 mm, 155 mm) prepared for each specimen, and further information added as this became available (fig. 4). This stored it in a way which made it compact and easily retrievable for subsequent sorting and analysis. Cards were coded directly to assist cross-

TABLE 1 Total number of animal specimens collected and examined listed according to major phylogenetic groups and periods of fieldwork

	1969-71[1]	1973	1975	1981	1986, 1990	Cumulative total
MAMMALS	33(20)[2]	2	8	-	-	43(20)
BIRDS	3(53)	-(48)	3(57)	-(6)	-(8)	6(172)
REPTILES	73(10)	-(11)	45(1)	-(2)	-(1)	118(25)
AMPHIBIANS	38(6)	-(6)	14(27)	-(1)	-	52(40)
FISH	-(31)	-(7)	-(6)	-(2)	2(40)	2(86)
Total vertebrates	147(120)	2(72)	70(91)	-(11)	2(49)	221(343)
CRUSTACEANS	-	-	3	-	7	10(0)
INSECTS	55(23)	22(-)	139(-)	2(-)	26(-)	244(23)
MOLLUSCS	17(7)	2	21	-	2	42(7)
OTHER INVERTEBRATES	-(26)	-(3)	3(32)	-	2	5(61)
Cumulative total	219(176)	26(75)	236(123)	2(11)	39(49)	522(434)

Notes: 1. The first figures in each column refer to actual specimens collected, preserved and deposited in reference collections. The figures in parentheses refer to all additional specimens (approximate numbers only) examined and then released, or discarded, or observed at close quarters in their natural habitat. 2. Mainly captured as game in traps or during hunting.

checking, reference and preliminary hypothesis-testing while still in the field. The main disadvantage of these cards is that they provide no duplicate record in the event of damage and are bulky, while preparation and sorting are both time-consuming and tedious. If I were embarking on this work now I would use a lap-top computer with back-up files on separate disks or cassettes, and enter data directly in the field.

Each specimen was identified individually with waterproof ink on parchment and the label attached physically to the specimen, or in the case of small invertebrates, placed in the glass phial (if wet), or written on the envelope (if dry). Eight fields of zoological information were recorded on the upper half of the front of each accompanying specimen card: (a) specimen number, (b) english name or appropriate taxonomic status, (c), scientific name (usually added after formal identification by receiving institution) (d) form and condition of preservation; or if discarded reason for doing so, (e) date and approximate time of collection, (f) location and elevation, (g) short description if to be discarded, or if features are likely to alter on preservation (e.g. coloration, critical measurements, life posture, live weight), and (h) means of acquisition (e.g. trapped, live gift). Additionally, five fields of ethnographic information were recorded on the lower half of the front of each card: (i) vernacular name of animal (if any), (j) name of informant, and (k) context of elicitation (e.g. ethnographers house, bush). Information i-k was recorded independently for as many individual informants as possible, and in most cases for at least more than one. In over 50 percent of cases I was also able to record responses to variations on the standard question frames [c.f. Berlin et al, 1974: 52]: (l) 'what is X a kind of?' and (m) 'how many kinds of X are there?', though I discuss in *The Cultural Relations of Classification* my reservations with such formulae. Any additional explanatory notes relating to data on the front of the card were placed on the reverse. Here also were written references to relevant data on this and related specimens documented elsewhere: on magnetic tape, film or in chronological field notebooks. Coding was accomplished in the usual way by punching combinations of holes along the edge of the card. Most of the analysis has been conducted on data in this form, although in 1987 all existing card records where transferred to a computer database using the same number of fields. This has permitted a final checking of the data which appear here in tabular form.

My Nuaulu work has involved a strategic combination of formal methods of elicitation and an observational and conversational approach in as near as possible natural settings. While I have always tried to be systematic wherever possible, I have also been blatantly opportunistic when I have felt that the occasion has merited it. I have experimented with more formal approaches using a relatively small number of informants in Rohua

FIGURE 4: Display of data on punched specimen cards used during field-work: *a* showing designated fields, and *b* illustrating a completed entry.

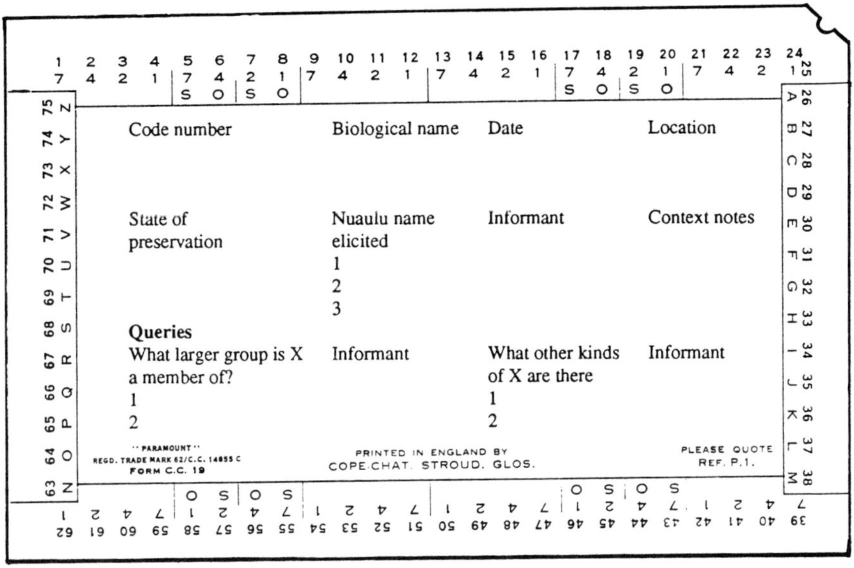

a

Code number   Biological name   Date   Location

State of preservation   Nuaulu name elicited 1 2 3   Informant   Context notes

Queries

What larger group is X a member of? 1 2   Informant   What other kinds of X are there 1 2   Informant

"PARAMOUNT"
REGD. TRADE MARK 62/C.C. 14855 C
FORM C.C. 19

PRINTED IN ENGLAND BY
COPE·CHAT. STROUD. GLOS.

PLEASE QUOTE
REF. P.1.

b

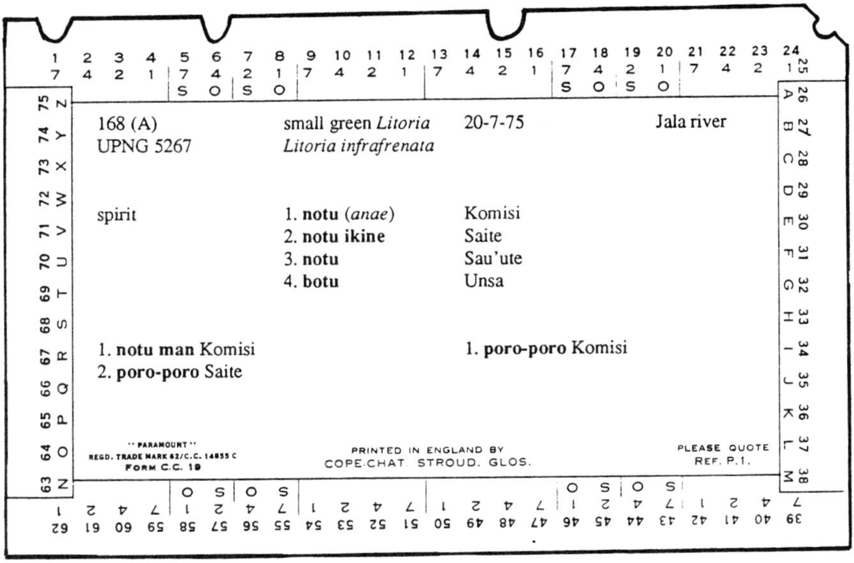

168 (A)
UPNG 5267   small green *Litoria*
Litoria infrafrenata   20-7-75   Jala river

spirit   1. **notu** (*anae*)   Komisi
2. **notu ikine**   Saite
3. **notu**   Sau'ute
4. **botu**   Unsa

1. **notu man** Komisi
2. **poro-poro** Saite   1. **poro-poro** Komisi

"PARAMOUNT"
REGD. TRADE MARK 62/C.C. 14855 C
FORM C.C. 19

PRINTED IN ENGLAND BY
COPE·CHAT. STROUD. GLOS.

PLEASE QUOTE
REF. P.1.

village, but have throughout put great emphasis on interpreting results contextually. Although systematic responses and specimens were obtained from only a relatively small number of Nuaulu informants, general ethnozoological data arose from interactions with numerous individuals in Watane, Bunara, Aihisuru and Hahuwalan, as well as Rohua. Data on the identification of specimens and discussions of classification and zoological knowledge were also accumulated on photographic film and sound-recording tape. The latter included conversations about animals, classification, behaviour, uses and general lore, verbatim records of identifying sessions, song, myths and stories, and other information relating to zoological knowledge. In some cases such conversations involved the ethnographer as a participant; others consist of long stretches of uncontrolled discourse about animals. Some recordings are of calls for different species made in natural habitats and imitation calls and decoys produced by Nuaulu informants. Permanent records of interviews were put on tape wherever this was physically possible. Information which could not be obtained during the examination of specimens or through general conversations with informants was committed to page-numbered duplicate notebooks in the usual way. As well as obtaining whole zoological specimens, artifacts manufactured from animal parts or associated with social and economic uses of animals were collected for subsequent analysis.

I found it useful to employ a number of simple introductory picture books [e.g. Tweedie, 1970] to stimulate talk and to compile initial lists. Even pictures of quite unfamiliar animals proved helpful in setting-up hypothetical tests to highlight significant criteria and configurations used to assign animals to particular categories and the extent to which particular categories could be manipulated. Thus, the rhinoceros hornbill of Malaya and the Sunda islands (*Buceros rhinoceros*), quite different from the species found of Seram was still, without any doubts, assigned the term **sopite** applied to the local hornbill (*Rhyticeros plicatus*). It was with more uncertainty that the Malayan tapir was labelled **maisan** (variously lion, tiger, elephant); none of which Nuaulu had ever seen.

In 1975 150x100 mm picture cards were prepared of animals with which the Nuaulu were known to be familiar from earlier fieldwork. Each card featured a clear and detailed illustration of one representative of a species reproduced from various studies of regional fauna (fig 5). This was accorded a code number (top right) and in the top left hand corner was placed the English gloss, scientific name, Nuaulu name, Ambonese Malay or standard Indonesian equivalent and the scientific name. The cards were divided into two series. Series 1 contained 42 cards with coloured illustrations of all the major common faunal types which I judged the Nuaulu to be familiar with. Series 2 contained 40 cards with mostly black-and-white illustrations of less familiar (but nevertheless common forms), for the most part consisting of invertebrates. The types varied from representatives of particular species known from the Nuaulu area (e.g. *Felis catus*, the domesticated cat) to generalised life-forms distinguished at the level of phylogenetic order or above (e.g. star-fish, butterfly). Like the picture-books, these cards served as useful stimulation for discussion and for informal and critical demonstrations of how Nuaulu principles of classification operated, with respect to hypothetical cases presented by unfamiliar species. They were also used as indicators of the kinds of specimens which the investigators were interested in collecting, and were employed on a limited and controlled basis in tests on more inclusive categories.

The drawback of using cards or other pictures as stimulus materials, and for tests, is in the use of representations of animals rather than the real thing, which means that occasionally the cultural images projected may be quite misleading. Pictures, for example, do not show motion (an often critical feature), colours are often 'unnatural', and animals are seldom the correct size, even relative to one another. In other words, the images have been decontextualised. Furthermore, the criteria used by informants are often not those employed by the informants themselves. Since the work of Heider, 1972 and others we have hopefully become more aware of the considerable problems of using card tests with unsophisticated and often illiterate informants.

I discuss in more detail the whole question of the relationship between techniques of elicitation and results in *The Cultural Relations of Classification*, particularly the use of formal question frames, formal tests of classifying behaviour, the role of the literate mode in both the offerings of those interrogated and in the interpretations of the interrogator, and the different ways of presenting data. It is to this work that those interested in pursuing these issues should turn.

1.5 The presentation of data in the present work

In recording repetitive systematic information I have tried to be as clear, concise and consistent as possible. To this end I have used a number of abbreviations, and these are listed separately at the end of the book. To facilitate location and cross-referencing of data, sections and sub-sections are indicated by a numerical code (e.g. 12.2.2.3), where the first numeral is the chapter, the second the section, the third the sub-section, and so on.

In each chapter information has wherever possible been given in the following sequence: (1) summary description of the relevant fauna for south

FIGURE 5: Examples of picture cards used in field sorting tests. Illustrations reproduced from M. Tweedie, *Animals of southern Asia*, by courtesy of Hamlyn Publishing Group Ltd.

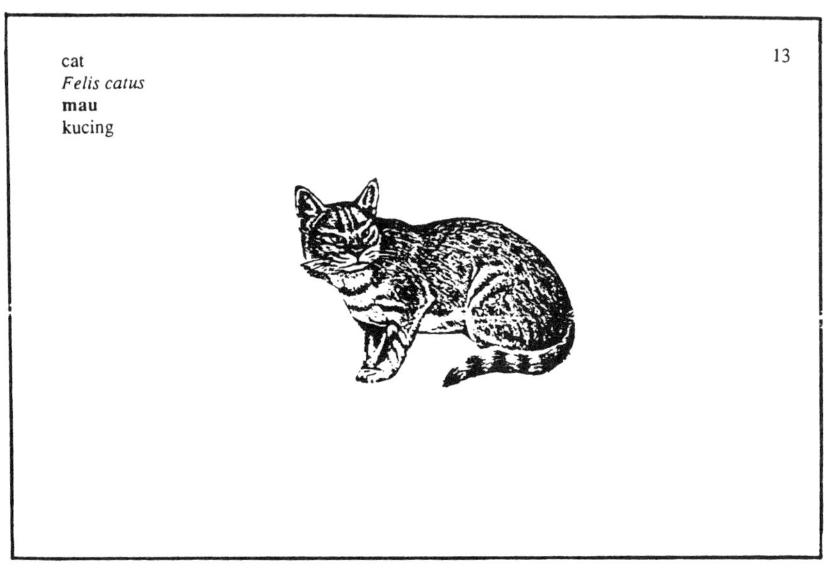

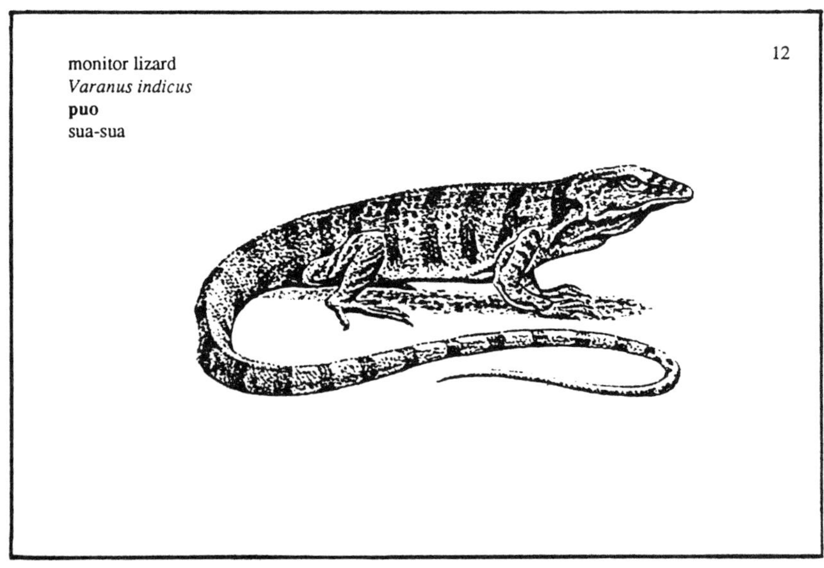

central Seram, (2) systematic listing of all Nuaulu categories applied to that fauna, (3) an analysis of the relations of non-basic Nuaulu categories, and (4) a brief account of social and economic uses. Where there is little matter to be included, some of these sections have been run together, as with sections (3) and (4) in chapter 14.

Each entry in section (2) begins with a list of terms applied to the category. Entire terms or part-terms appearing in round brackets indicate instances of free variation. Synonyms simply follow on from each other, separated only by a comma. In subsequent references to the category, and including all tables, the citation form chosen is what I judge to be the most commonly found lexical variant of the most commonly used synonym. Sometimes this decision has perforce had to be a fairly arbitrary one.

Each description generally begins with a semantic analysis of the term. If a term is not further reducible semantically, and if there are no clues as to its meaning or etymology, this is indicated only by omission. This is followed by information on the phylogenetic species to which the terms are applied, and occasionally their Ambonese Malay or Indonesian gloss. The entry concludes with a discussion of the cultural significance of the category.

I have used Linnaean taxonomic terminology only in so far as this is minimally sufficient to identify the range of categories and place species in the broader phylogenetic context. I have used the following typefaces (upper case Roman, italicised capitals, lower case Roman, italicised lower case) to indicate the different levels of taxonomic organisation:

PHYLUM	*ORDER*	superfamily	*genus*
CLASS	*SUBORDER*	family	*species*
SUBCLASS		subfamily	*subspecies*

My knowledge of Nuaulu ethnozoology is more thorough for some groups than others. Thus, I am particularly aware of the shortcomings in my account of bird categories, which arise from the inability or unwillingness to obtain sufficient live or preserved specimens for the purposes of identification. In view of these problems, I provide discussion on informant variation only where my data are most complete.

Finally, in their various publications on Kalam ethnozoology, Bulmer and his associates [e.g. Bulmer and Tyler 1968: 381, Bulmer and Menzies 1972-3: 102] have frequently presented charts plotting number of actual specimens collected against local terms (horizontally) and biological names (vertically). Bulmer's own procedure [personal communication] has in most cases been:

(a) to list the identifications provided by the original captor of a speci-men, on the grounds that he has the greatest range of circumstantial information at his disposal; except

(b) when several persons are present at the place of capture, when the consensus identification is accepted for tabulation. If there is no con-sensus the identification accepted is that of the person reckoned to be most competent of those present.

Obviously, this only works if (a) identifications of single specimens are con-sistently unambiguous or (b) if a decision is taken to accept one identification as the 'correct' one. Many of the specimens collected on Seram were 'identified' by more than one informant, sometimes up to six. Frequently the response of one informant would contradict that of another. Consequently, the numerals in all but the initial (N) column of tables of species identifications presented here represent the number of individual responses.

Notes to Chapter 1

1   Algemeen Rijksarchief, The Hague: VOC 1293 (f.176r, 1.84r), 1317 (f. 200r-v, f.71), 1.344 (f. 56), 1368 (f. 40v) and 11245 [Anon., 1908: 130-31].

2   The most regular orthographic usage adopted by the Indonesian administration. In previous publications I have written 'Ruhuwa', in order to harmonize with the rendering found in some earlier Dutch maps (e.g. Topographische Inrichting 1919, *Schetskaart van Ceram* (Scale 1 : 100, 000), Batavia).

3   See Ellen, 1985, also for certain nomenclatural revisions to Ellen, 1973: 140-8; Ellen, 1975b and; Ellen, 1978c: 65-8. Further information relat-ing to other matters discussed in this section may be found in Ellen, 1973: 20-63, 368-73, 391-2, 446-64; and; Ellen, 1978c: 212-9.

4   For further details on the history of the systematic zoology of the area see Nolthenius, 1935; Rubenkoning, 1959:ii-iv; Ruinen, 1928.

5   For convenience, the ethnographic present should be understood as 1970-75. Where data collected on subsequent occasions might reason-ably be expected to reflect recent changes I have tried to make this clear in the text.

6   The term *phylogenetic* is used throughout this work to cover all forms of 'scientific' taxonomic, systematical and nomenclatural practices. The term is imperfect, but I find the possible alternatives - 'Linnaean', 'Western', 'scientific', and so on which I am occasionally tempted to resort to - equally if not more problematic.

CHAPTER TWO

TERRESTRIAL MAMMALS

2.1 Terrestrial mammal fauna of south central Seram

The available information on terrestrial mammals in the Moluccas (excluding bats) reflects the generally depauperate fauna. 25 certain species are known from Seram, including about 15 introduced or domesticated by *Homo sapiens*. Of these, 14 were observed during the course of fieldwork in the Nuaulu area. Because of the larger size of land mammals, relatively few specimens were collected compared with other groups, apart from birds and fishes. Complete specimens were restricted to smaller species, although bone material from larger species was collected when available.

Those certain species reported from Seram, but not observed or collected include the following:

Peramelidae

    Rhynchomeles prattorum, Seram island bandicoot

Muridae

    Murinae

    Rattus ceramicus

    Rattus rattus manuselae, house rat (black rat)

    Rattus ruber feliceus

    Melomys aerosus

    Melomys fulgens fulgens

    Melomys fraterculus

    Nesoromys ceramicus

The difference between the number of species observed and those known zoologically from the entire island may be accounted for by relative geographical distribution, particularly differences between highlands and the coast, and the smaller population sizes of unobserved species. For example, all species listed above, with the exception of *Melomys fulgens*, are known only from specimens obtained in the Gunung Binaiya area, between 1000 and 2000 meters.

A checklist of the land mammals reported from south central Seram is presented in table 2.

TABLE 2 Checklist of terrestrial mammal fauna (excluding bats) recorded in the Nuaulu region of south central Seram.

Species	Ecological Zones 1 2 3 4				Nuaulu glosses
MARSUPIALIA					
Phalangeridae Phalangerinae					
Phalanger orientalis orientalis cuscus	+	+	+	-	**mara kokowe (♂)** **mara osu (♀)**
Phalanger maculatus *chrysorrhous*, spotted cuscus	+	+	+	-	**mara makinete (♂)** **mara siha (♀)**
INSECTIVORA					
Soricidae					
Suncus murinus murinus house shrew	-	-	+	-	**kusa-kusa**
RODENTIA					
Muridae Murinae					
Rattus rattus rattus house rat; black rat	-	-	+	-	**imanona**
Rattus exulans ephippium Pacific rat	-	-	+	-	**imanona, mnaha**
Mus musculus musculus common mouse; house mouse	-	-	+	-	**mnaha**
Melomys mosaic-tailed rat	-	+	+	-	**imanona ai ukune**

CARNIVORA

Canidae

Canis familiaris
domestic dog   - - + -   asu

Viverridae
Viverrinae

Viverra tangalunga tangalunga
the Malay civet   + + + -   lau

Paradoxurinae

*Paradoxurus hermaphroditus
setosus*, common palm civet + + + -   kuha

Felidae

Felis catus
domestic cat   - - + -   mau

PERISSODACTYLA

Equidae

Equus caballus
horse   - - + -   naitanane

ARTIODACTYLA

Suidae

Sus scrofa   + + + -   hahu
Sus celebensis   + + + -   hahu numte
wild pig

Sus domesticus
domestic pig   - - + -   hahu pale

Cervidae

Cervus timorensis moluccensis   +   +   -   **maianane**

Bovidae

Bos Taurus/indicus
domesticated cattle   -   -   +   -   **korobou, sapi**

Capra hircus
domesticated goat   -   -   +   -   **une-une**

Key: Zone 1 = above 1000 metres, principally montane rain forest; Zone 2 = tropical rain forest; Zone 3 = secondary forest, garden and village areas; Zone 4 = freshwater and swamp forest.

2.2 Nuaulu categories applied to land mammals

2.2.1 marane

The category **marane** corresponds to the genus *Phalanger,*[1] in terms of morpho-syntactic status of nomenclature, classification and semantic reference of the label. It is directly equivalent to AM 'kusu'. Morphologically, behaviourally and ritually it cannot readily be confused with any other animal with which the Nuaulu are familiar. Although the distinction between placental and marsupial is of no significance in Nuaulu ethnozoology, it does not follow that the Nuaulu fail to recognise the difference. It is omnivorous (eating fruits, insects and sometimes small vertebrates) and truly arboreal. In fact, apart from certain introduced domesticated species generally not kept by the Nuaulu (deer, pigs and goats), and perhaps civets, it is the largest mammal and quadruped known to the Nuaulu. In addition, it can be readily distinguished aurally, by smell, and by the taste of the meat.

Although discrimination of cuscus from non-cuscus is a matter for little complication, the internal ordering of the category **marane** is not so clear-cut [see Ellen 1975a: 204, table 1, and related text]. It comprises four terminal categories, corresponding to two quite distinct sexually dimorphic species:[2]

2.2.1.1 **mara kokowe** The term is prohibited in the company of the opposite sex, as it stands metaphorically for the penis. In such circumstances the synonym **mara hanaie** (lit. male **marane**) is used. Black dorsal line present, grey; greyish-brown to white; throat to neck suffused rufous **msinae** in breeding season: male *Phalanger orientalis orientalis*.

2.2.1.2 **mara osu** Possibly a contraction of **osu nakatu** (cockatoo) or **mara osune**. Black dorsal line present, smaller and darker than **kokowe**; similar colouring but no rufous suffusion: female *Phalanger orientalis orientalis*.

2.2.1.3 **mara makinete** Sometimes contracted to **mara inete**. No dorsal line; white body usually spotted but considerable variation in colour; largest cuscus category: female *Phalanger maculatus chrysorrhous* (plate 7).

2.2.1.4 **mara siha** No dorsal line; darkish body usually spotted; dark flanks in contrast to white belly: female *Phalanger maculatus chrysorrhous*.

Generally, informants felt that accurate identification and description required the combination of a number of different types of feature, although standards of relative size and colour were spontaneously offered, but with the caveat that such keys were not a consistently reliable means of identification. Thus, for colour in order of increasing lightness the categories were ordered **osu-kokowe-siha-makinete**; the series also appearing in the same order in terms of increasing size. The circumspection with which such keys were offered indicates a general sensitivity to variation in the distinctive features of cuscus categories. Indeed, except for distinctions such as dorsal line present/dorsal line absent, male genitalia/female genitalia, it would prove exceedingly difficult to describe the differences between cuscus categories purely in terms of contrastive features. In particular, it was stressed that **mara makinete** above all others varied widely in coloration, ranging through black, grey, reddish-brown and white. That such variation is sometimes attributable to age is recognised by Nuaulu informants. This is particularly marked in **mara kokowe,** where there appears to be a consistent terminological distinction between younger forms (or **mara koko putie**) and older forms (**mara koko msinae**), where the younger are distinguished by their generally much whiter and lighter coloration, and the older by a yellow-brown breast. **Mara kokowe** of an intermediate hue are sometimes called **mara makioi,** [3] referring to their mixed (reddish-brown, yellow-brown) coloration. Occasionally animals of indeterminate categories are labelled provisionally as **mara putie** or **mara metene,** referring respectively to specimens of a white or darkish hue, while specimens of **mara kokowe** and **mara siha** showing a rufous suffusion (typical of the breeding season) are at times termed **mara hehue.** The list could be extended, but but as it stands amply demonstrates that despite the plethora of terms, there is no

PLATE 7: Spotted cuscus, probably male (**mara makinete:***Phalanger (Spi-locuscus) maculatus*). Photograph reproduced by courtesy of the Zoological Society of London. Print no. 711.

question of any of them being considered as 'natural kinds', so much as descriptions of variant individuals. Predictably, it is the younger immature animals that are the most difficult to identify and irrespective of category these are called **mara porune,** for which restrictions appear not to be so rigidly enforced[4]. In general though, discrimination between adults of these different categories is regarded as being straightforward and Nuaulu are adept at making rapid visual and auditory distinctions.

I have elsewhere [Ellen, 1975a: 207-11] demonstrated how cuscus classification varies according to a combination of one of several contexts with different individuals and/or clans. These contexts are: occasional chance observations or hearing in the forest while engaged in other activities, confrontation while engaged in hunting cuscus or while consuming its meat, in general conversation, and in ritual. Different situations yield three alternative means of classification: *lexical* (or morpho-syntactic), *ritual* and *morphological*.

The lexical classification of cuscus categories is merely an undifferentiated list, an *index* [Conklin 1964: 40], the contents of which, for the purpose of listing, are taken as having no further criteria of distinction other than their common membership of the category **marane.** This form of primary classification is based on simple class-inclusion and such an undifferentiated form of listing is often the kind of response which an ethnographer first elicits. Certainly he or she cannot afford to assume otherwise at an initial phase in an investigation, although (as has been pointed out by Heider, 1972: 465, n. 1), the bias which can follow from assumptions to the contrary is widespread in ethnographic interpretation. In such a classification the order in which the terminal categories are given is irrelevant, or must initially be taken as being so. Thus, this was the first arrangement for the category **marane** which I elicited in the field from informants during formal sessions of questioning. Evidence obtained subsequently suggests that this kind of arrangement is that most commonly used, or rather was that minimally necessary and adequate for most circumstances in which the cuscus is involved. In eliciting, checking and counter-checking the terminal categories in this broader category, I was presented with the items in varying orders. At the time such consistent inconsistency suggested that for the purpose of listing order was an immaterial consideration. It should be noted that it is only this arrangement which is actually realised in the morpho-syntactic structure of the category, and it is for this reason that I have labelled it the *lexical* classification.

Cuscus is important in a number of major ritual contexts: male initiations, birth rituals, the great ritual house festival, marriage payments and first fruits ceremonies. But it is only for those clans where **mara makinete** is totemically salient, that we can speak of a *ritual* arrangement, in which this category is contrasted with the remaining three. **Mara makinete** is a primary totem for three, and a secondary totem for four, of the existing ten Nuaulu clans [Ellen 1993: table 6.3]. Thus members of all such clans consider it to be **monne,** sacred; the consumption of its flesh and slaughter prohibited. Finally. there are two alternative *morphological* arrangements: one based on the grouping of like sexes, one based on the grouping together of members of the same species into an unlabelled 'natural kind'.

From the foregoing it will be evident that cuscus is a significant source of animal protein, in fact the third most important in terms of gram-weight consumed[5]. Cuscus are very occasionally kept as household pets, though they tend to be rather smelly. The half-mandibles of butchered animals are used as borers and gravers.

2.2.2 lau, kuha

These two closely-related, and probably overlapping, categories refer focally and respectively to the civets *Viverra tangalunga* and *Paradoxurus hermaphroditus;* a conclusion based on descriptions given to me of size, behaviour, coloration and general morphological characteristics.[6] The Malay civet, *V. tangalunga,* has not been previously reported from Seram. It is known from Ambon and Buru, to where it was presumably imported for its musk, used as a perfume fixative [see e.g. Gijsels, 1871: 386-7]. Its existence might reasonably be suspected and I am satisfied that evidence collected by me during fieldwork confirms this, bar the provision of actual specimens. *P. hermaphroditus* has probably been present for many millenia and a date of 4000-5000 BP is reported for Timor [Glover, 1986: 159]. **Kuha** is said by some informants to be a type of **lau,** but it is likely that both **lau** and **kuha** are seen as being sufficiently closely related that either might be said to be a type of the other. The fact that there is a great deal of individual and racial variation in the coloration of *P. hermaphroditus* [Pocock, 1933: 014] is likely to increase difficulties in identification.

Civets are only rarely eaten, and **kuha** in particular is considered difficult to hunt. As some young men in Rohua have never seen a **lau** or **kuha** at all, it is perhaps not surprising that I did not manage to obtain any specimens. **Lau** is described as living on the ground with grey/yellow coloration, and with strongly marked black and white tail rings. **Kuha** is described as arboreal. The AM glosses 'tinggalam' and 'musang' (sometimes 'kesturi') are employed to refer respectively to these civets, and the distinction also finds support in the equivalence of AM 'musang' with **kuha** in the language of Piliana in the central highlands. Both civets are said to prey upon cuscus, a matter of some practical concern to Nuaulu.

2.2.3 tui-tui

Tui-tui (for which **makueni** is the more rarely heard perfect synonym) is also glossed as 'musang' by some Nuaulu informants, and is almost certainly onomatopoeic. It is compared with **lau** and **kuha**, and said to be a relatively common animal of the forest floor, reputedly eating (amongst other things) adders and skinks. It is regarded by Nuaulu as stupid as it always comes back to defecate in the same place, and it is this which may account for relative hunting success. In the course of a four month dietary survey I recorded the consumption of 0.44 grams per adult head a day of what was described as **tui-tui,** that is about one percent of the total amount of animal protein consumed. However, I believe this figure to be rather untypical.

Descriptions of this creature certainly suggest affinities with viverrids, though the existence of the term **tui-tui** in the language of Piliana, in addition to terms which gloss the Nuaulu categories **lau** and **kuha,** is evidence for its status as a separate and contrasted form. Informants described **tui-tui** as being similar to **lau,** with large teeth, tail like a cat and fine hair. I retain an open mind as to the identity of this small mammal. Its significance in Nuaulu thought is attested to by a reference in an **auwoti** (war dance) stanza comparing the Nuaulu with the **tui-tui,** neither of which should ever leave the mountains:

Tui-tui helete   The tui-tui descends from the mountains

Helete Nunusaku   Descends from Nunusaku

From this it is tempting to conclude that it might be the marsupial bandicoot *Rynchomeles prattorum.* An alternative possibility is a large forest rat, though this would seem a disappointingly pedestrian candidate.

2.2.4 hahu

The term is used polysemously to refer to Suidae in general, and more specifically *Sus.* There is considerable variation amongst the common wild pigs of the Moluccas, ranging from *S. scrofa* to *S. celebensis,* the former having been dispersed early as a domesticated type and later interbred with *celebensis* [Groves, 1985: 433-6]. All pigs of the *Scrofa* group appear to be feral descendents of individuals carried from island to island by human agency, although recent carbon 14 dates suggest an early appearance for parts of eastern Indonesia. A date of 5,520+ 60 BP has been recorded for a site in Timor [Glover, 1971: 176]. Recently introduced domestic breeds (**hahu pale:** *Sus domesticus*) are known from Christian settlements.

In its wider sense **hahu** (plate 8) also includes the categories **hahu numte** and **hahu nihu**. The first of these, distinguished by the Nuaulu by its larger tusks and wartiness, probably refers to pigs of the *Sus celebensis* group. The second is said to have existed on Seram in former times and, though mythical (since the Nuaulu have no evidence that it did once exist on Seram), it may be associated in Nuaulu thought with the babirusa, *Babyroussa babyroussa,* known from the nearby island of Buru.

Wild pig is by far the most important source of animal protein in Nuaulu diet, constituting some 30 percent of the total weight consumed. It is hunted by individuals or groups, with spear, bow-and-arrow, trap and - when available - breech loading rifles. Pork may be brought back to the village either raw or ready-roasted. Tails (**hahu etute**) are removed when animals are butchered, attached to a length of rattan or thread and hung as a charm round the necks of young children to ward off ill health [Ellen, 1993: plate 1.6h]

My impression is of a high density of wild pig in Seramese forests, due to only slight pressure through hunting, restricted as it is to animist and - to a lesser extent - Christian groups. In this respect, the irregular migration of pigs in search of fruits is not of particular importance to Nuaulu hunters, although they are aware of it. It is well known that certain places are rich in potential pig foods. Hunters have a thorough knowledge of foods eaten by wild pigs, and use this to their advantage. The fruits of *Canarium indicum* and other *Canariums, Terminalia catappa, Jatropha curcas, Quercus molucca* and *Ficus altissima,* and the leaves of *Adenostemma lavenia* and *Laportea decumana* are said to be particularly favoured. They are also said to eat the faeces of cassowaries, though I have no information which might suggest that it is considered a delicacy.

Pigs often raid gardens, rooting around for tubers (especially taro and sweet potato) and fruits. Their deleterious effects may be countered by traps, but fencing is not employed widely. The Nuaulu occasionally fatten pigs, but there is no tradition of domestication and breeding. Of the few cases that I observed in the field, piglets had been acquired live when their sows had been trapped or caught in a hunt. Often, such pigs are slaughtered before they are mature, due to their high nuisance rating. They are not penned, but are allowed to roam free and forage on vegetation and domestic growth in the village area or on its periphery. Sometimes they are fed on the refuse produced in the processing of sago. On Seram, domesticated pigs are only kept in Christian villages, where they may be penned, systematically fed and bred, or allowed to roam free.

PLATE 8: Tame pig (**hahu**:*Sus scrofa*)being fed on coconut waste, Rohua: 1 February 1981 (neg. 81-4-10a).

PLATE 9: Resting on boulders of a dry river-bed while hunting with dogs: 1 April 1970 (neg. 70-4-5).

The general ritual and symbolic significance of pigs is linked to their membership of the category **peni,** discussed in chapter 4 of the *Cultural Relations of Classification*, and briefly below (2.3). In addition, because of the Islamic proscription on pork consumption, the close proximity of Nuaulu to Muslim settlements, and because many Moluccan Muslims fear ritual contamination from anything connected with pigs, the pig has become a focal symbol of Nuaulu identity. Pig grease is rubbed over the bodies of men for certain major rituals, a practice which serves to accentuate this latter purpose.

2.2.5 maianane

This term is used unambiguously to refer to *Cervus timorensis moluccensis*, the Moluccan deer, but may also be employed to cover other breeds of deer that have been introduced into Seram. Nuaulu distinguish between small deer with immature antlers (**maianane tanapaku**) and older deer with large and mature antlers (**maianane mainihune**). **Maianane tanapaku** generally refers to the first pair of antlers, consisting of irregular and asymmetric single branches without ramifications, but may also be extended to the second pair which may also consist of single branches, but which normally have four points (2+2). **Maianane mainihune** usually refers to animals with the third pair of antlers and invariably six points (3+3).

Wild deer is the fourth most important source of animal protein in Nuaulu diet, constituting some 12 percent of the total weight consumed. Like pig, it may be hunted by individuals or groups, though it is generally caught by single hunters. The same weapons are used in hunting deer as in hunting pigs, and the meat is prepared and cooked in a similar way. The deer population of Seram is large and individuals and herds up to 20 strong are commonly found brousing on shrubs and herbs in coastal groveland secondary growth.

The deer is not domesticated, although occasionally young animals may be captured and sold to non-Nuaulu as pets or as meat. It has been suggested by Wallace, 1962 (1869): 300 that the deer was originally introduced into Seram by Muslims, who still keep the animals and who consider the meat a great delicacy. Presumably, the wild stock originated from tame deer which were deliberately allowed to run feral or which escaped from captivity. Indeed, importation into islands previously uninhabited by deer appears to have taken place frequently in the archipelago. For example, *Cervus timorensis moluccensis* has been imported into the Obi group (1930), the Aru islands (from Seram, 1855) and the western part of the Onin peninsula of New Guinea (from Seram, 1913). Following Valentijn, 1724-26, Van Bemmel [1949] suggests that the deer of Amboina are the descendents of seventeenth century imports of *Cervus timorensis rusa* from Java and *Cervus timorensis macassaricus* from Sulawesi. There are no known archaeological remains of deer from the Moluccas, although prehistoric remains dating from the last 700 years are known from Timor [Glover, 1986] and from Sulawesi after 3500 BP [Glover, personal comm.].

Like pigs, deer are part of the wider Nuaulu category **peni**, from which they derive much of their ritual significance; that is, not as deer specifically, but as **peni**. Non-food uses of deer include the making of drumskins (**tihane unte**) from the hide (e.g. B.M. AS. 1.177 and Ellen 1970.617), the use of young antlers either singly (B.M. AS. 1.9) or in pairs attached to the cranium as househooks (Ellen 1993: plate 1.6b). These latter are known as **tanapaku** and the term for young deer - **maianane tanapaku** - would appear to be derived from this usage, since **paku** are also pins or pegs, particularly the wooden variety used for construction purposes. Large antlers (**sepi-sepie**) are sometimes sold and are an ingredient in certain Chinese medicines. Gimlette, 1971 [1915]: 56 reports that deer antler may be an ingredient in the magical prescriptions of peninsular Malays, but I have come across no similar applications in the Moluccas. The white fur and attached hide from the back of the ear is used to make ear ornaments (**maianane tina totue**) worn by men (e.g. B.M. As.1 233).

2.2.6 asu

This term is used unambiguously to refer to the dog, *Canis familiaris*. It is fully domesticated on Seram, although bitches give birth to their litters in nearby forest and puppies generally have to be brought back to the village if they are to survive or not run wild. There are some wild dogs (**asu manene**), which have probably escaped from domestic stock. Domesticated dog is known archaeologically elsewhere in Indonesia from around 3000 BP (e.g. from Timor [Glover, 1986: 205]).

The Nuaulu have a further sub-category of dog, **asu nau,** which is contrasted with ordinary domestic dogs. This was described to me as a large 'forest dog' which once existed but which is now extinct. **Nau** is 'divination' and **nau one** means 'the beginning', generally 'the beginning of the world', 'the creation'.

Domestic dogs are used primarily for hunting (plate 9), but they also serve to guard dwellings and act as scavengers of household waste. They will occasionally eat small vermin. In hunting they may be used singly or in groups. In Rohua each household had an average of three dogs, good hunting dogs being selected as young puppies, and named. Weak, diseased and imperfect dogs are weeded out through sheer starvation. Dogs cannot be killed, except on payment of a fine, and even a dog that has been wounded in the chase will be left to die a 'natural' (that is a *good*) death. Occasionally, hunting dogs which are diseased or wounded will be given treatment, the knowledge of which is vested in particular individuals and clans. An annotated abstract from my fieldnotes [1975-2-37] underscores Nuaulu attitudes to dogs well:

> Numapena mentioned to me today that he has three scruffy dogs (**asu hoinc:** 'eczema dogs') - no doubt the rump of the litter - which have been biting his children. He is anxious to dispose of them. With some trepidation, and after turning the matter over in his mind for some days, he has now killed them. He took the carcases to the dry river bed of the Yoko, where he discarded them. Having done this he made three models of dogs from sago leafstalk (**tope**). These he took into the forest and set on poles near the spot where he had thrown the dogs, as if they were **wate** (scare charms). These, he says, are **asu siaie,** and he performed the magical **ye ruhu asu** procedure to assuage the anger of Anahatana and the spirits of the dogs.

I was told that dog may occasionally be eaten, but I suspect that this is only as a last resort, or on special occasions, and then only with special permission from the ancestors. I never witnessed its consumption during fieldwork, a situation which contrasts markedly with the Ambonese. Preference is said to be given to forest dogs.

2.2.7 mau

This term, clearly onomatopoeic, refers to the domesticated cat, *Felis domesticus,* in particular the introduced European variety with the well-known (at least in southeast Asia) 'knotty-tail'. It is occasionally found in Nuaulu villages, where it is used to keep down vermin. It is not eaten, although to do so is not expressly forbidden.

2.2.8 une-une

The term refers to the domesticated goat, *Capra hircus.* The reduplicated form suggests an onomatope. It is possible that goats were only introduced to Seram in any number with Islam. On the other hand, Glover [1986: 205] has reported its regular occurrence on Timor from 3500 BP. Goats were present in Java by the mid-ninth century AD, probably of western Asiatic origin.

Goats were never found in Nuaulu villages until the late nineteen-eighties, and in fact their consumption and rearing is regarded as **monne** (forbidden). This is undoubtedly linked to their strong identity with Islam providing a means of symbolically relating to the village of Sepa which for many years has politically dominated the Nuaulu clans of south Seram. Goats are kept in small flocks in all Muslim settlements, where its meat is consumed on special occasions and is the supreme animal of sacrifice. Seen this way, the contrast *pig : goat* is part of a wider symbolic opposition between Sepa (standing for Islam generally) and the Nuaulu [Ellen, 1988b].

2.2.9 naitanane

This term refers to horses of all breeds (*Equus caballus*), whether introduced by Europeans from European stock, or whether derived from elsewhere. Valentijn reports 50 to 60 horses in Amboina in the latter part of the seventeenth century, many of them imported Javanese saddle-horses used for Dutch administrators and by local worthies, some of which were imported by Javanese and Macassarese [Kraneveld, 1959: 147]. Considerable numbers were introduced at the end of the last century by the Dutch, for military and administrative purposes, and large numbers of coastal and some mountain paths were specifically designated *paardenpad* or 'horse paths' (see e.g. sheet VIII of the 1922 1.100,0 map of Seram, *Topografische Inrichting*, Batavia).

The term **naitanane** is not obviously a recent introduction and it is possible that horses were introduced into Seram, and more generally to the Moluccas, before the second half of the nineteenth century. At the present time there are a small number of ponies in south Seram owned by non-Nuaulu peasants who use them for transporting, in particular, copra. Official statistics [Statistik Tahunan, 1974: 38] claim that between 1971 and 1973 there were no horses in the entire Central Moluccas. Although this is certainly incorrect, the numbers are clearly very small.

Horses are not, and have never been, kept or eaten by the Nuaulu, although consumption is not forbidden.

2.2.10 sapi

This term, from the Malay, refers to domestic cattle, *Bos taurus indicus*. Cows are now kept in small numbers by a few Nuaulu households (plate 10), and by other people on Seram. In 1973 the total number of cattle in the central Moluccas was reported at 11,494 head, having almost doubled in the preceding three years [Statistik Tahunan, 1974: 38]. In parts of the Moluccas (especially Ambon, Lease and Banda) cattle have been imported since the seventeenth century, although little provision has ever been made for the local increase of stock. By as early as 1663 there appears to have been uncontrolled breeding among escaped stock, which were then the object of institutionalised hunting [Kraneveld, 1959: 148].

PLATE 10: Cows belonging to Komisi Somori, Rohua: 26 July 1975 (neg. 75-2-19).

The Nuaulu may consume beef, but I have not yet witnessed a slaughter. It is likely that they are far more important as a source of cash. They are not used for milk. The stock seems to be from recently introduced strains under government supervision. Stock-raising is being encouraged by the provincial administration.

2.2.11 korobou

This term is now used interchangeably with **sapi** to refer to domestic cattle and appears to be the older and more frequently employed of the two. Like **sapi** it is derived from AM (in this instance, 'kerbau', meaning 'water buffalo', *Bos bubalus*), evident from the voiced stop 'b' not found in contemporary Nuaulu. Official statistics [Statistik Tahunan, 1974: 38] list no water buffalo for the central Moluccas, although there are large numbers in the southeastern islands. Buffalo appear to have been introduced into Ambon by the first part of the seventeenth century, where they were owned by local rulers [Kraneveld, 1959: 146].

2.2.12 mnaha(ne)

Also known as **mnaha niane** ('village') and **mnaha numa** ('house').

Out of 22 responses, all informants gave the term with reference to specimens of *Mus musculus;* four out of 10 identifications of *Rattus exulans* gave this label. All the evidence suggests that *Mus musculus,* the common commensal mouse introduced and distributed by human agency, is the focal content of this category.

At this point a note should be made concerning *Suncus murinus,* the house shrew. I have elicited no term for this shrew from the Nuaulu area, and collected no specimens. In the Malay peninsula, where it is known as 'cencurut rumah' ('house shrew'), it is confined to towns and suburban areas [Gathorne-Hardy (Lord Medway), 1978: 3] and is absent from truly rural habitats. It may therefore only be found on Seram in a few of the larger coastal settlements. In the animal terminology of peninsular Malays one form of shrew, *Crocidura fuliginosa,* is known alternatively as 'cencurut hutan' ('forest shrew') and 'tikus pahit' ('bitter mouse/rat'), suggesting their perceived close relationship and perhaps occasional conflation of shrews and mice. It is possible that the Nuaulu might, without difficulty, include it within the category **mnaha,** especially if it is called 'tikus' in AM; but see also 2.2.16.

2.2.13 imanona

Six out of 10 informant responses indicated the focal content of this category as *Rattus exulans.* Four specimens of *Rattus* sp. were also identified by the term. The category also probably includes *Rattus rattus,* a form introduced by human agency, *Rattus ruber,* and sometimes *Mus musculus* (see 2.2.17). By and large, though, it refers to forest forms living in the village and garden area. It is commonly associated with banana trees. *Rattus norvegicus* is generally restricted to large coastal settlements, and may not be known to the Nuaulu.

2.2.14 imanona ai ukune

No specimens collected or identified, but the adjectival qualifier (**ai ukune** = 'tree top', 'far forest') and informant descriptions of a brightly-coloured ochraceous tree-living animal, with a long naked tail, suggest *Melomys fulgens,* or possibly *M. aerosus, M. fraterculus, Nesoromys ceramicus,* or any combination of these four related species. It is said to often occupy the same trees as the fruit bat *Pteropus melanopogon* (**nota sapane**).

I have a single report of **imanona ai ukune** having been eaten.

2.2.15 mirine

This ground-living murid, known in AM as 'kuning' (yellow) is described as having a white underside, dark brown fur yellowing along the sides, a brown-red tail, and as being similar to **imanona**. Some informants say that it is a type of **imanona**. It is eaten by **tui-tui,** monitor lizards and snakes. In turn, it is said to eat crickets, bananas and leaves of forest trees, but not tubers and other common garden crops.

Mirine is possibly a synonym for **imanona ai ukune** (the latter being simply descriptive), in which case its content is identical. Alternatively, it may represent a contrasting species of *Melomys,* or perhaps *Nesoromys.*

2.2.16 kusa-kusa

This term is possibly of AM origin, and in peninsular Malay 'kusa-kusa' refers to grass-fodder supplied to domesticated animals. For the Nuaulu it refers to a rarely-seen murid or perhaps the shrew *Suncus murinus.* It is said that if a **kusa-kusa** asks for fire, you must give it to him.

2.2.17 The relationship between various categories for murids

Mnaha and **imanona** appear to represent two polar types within a broader semi-covert category. Different rats and mice are placed along a scale **mnaha** <--------> **imanona,** and their designation as either depends on the informants view as to whether an animal is 'more **mnaha**' or 'more **imanona**'. Such an interpretation is supported by informant variation in applying terms to specimens, as indicated in table 3.

Size seems generally unimportant as a distinguishing feature, although some informants claimed that **imanona** was as a rule larger than **mnaha**. On this basis, immature specimens of what might otherwise be considered as

TABLE 3 Species identifications compared with Nuaulu categories applied to 31 murid specimens

	unidentified rat	*Mus musculus*	*Rattus exulans ephippium*	*Rattus* sp. (prob *rattus*)	Total
mnaha (ne)	1	21	2		24
mnaha niane/numa		1	2		3
imanona	2		6	3	11
Number of informant responses	3	22	10	3	38
Number of specimens	2	22	6	1	31
No identification provided				1	1

imanona may be labelled **mnaha.** Many of the other morphological features appear to be unimportant, and the major criterion of distinction appears to be habitat. Thus **imanona** was sometimes described as **mnaha wesie.** However, it seems that the focal characteristics of the respective categories are reasonably definite, such that **imanona ai ukune, mirine** and **kusa-kusa** were sometimes described as 'types of' **imanona.** To a certain extent, this picture resembles peninsular Malay use of 'tikus' for all kinds of murids: *Mus* then becomes 'tikus terkecil' and *Rattus rattus,* 'tikus rumah'; *Rattus exulans,* 'tikus kecil' and *R. norvegicus,* 'tikus mondok' [Gathorne-Hardy (Lord Medway), 1978].

Looked at taxonomically, the more inclusive category may be occasionally verbalised as **mnaha** and occasionally as **imanona,** depending on context:

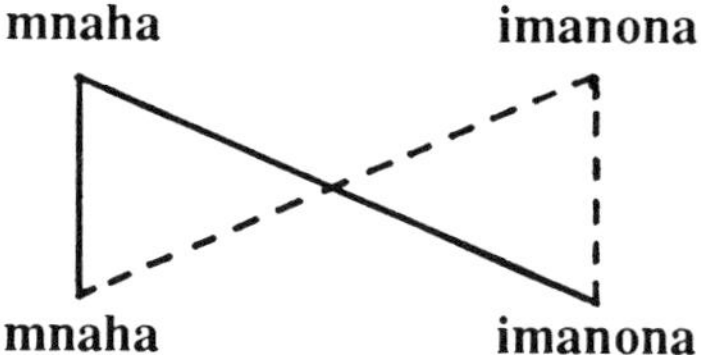

In its more specific sense, that is in contrast to **mnaha, imanona** may be partitioned into two further categories: **imanona ai ukune** (perhaps including all other forest rats) and residual **imanona**, ordinary village rats.

2.2.18 ruka

Glossed in AM as 'kera', a term which in peninsular Malay applies specifically to *Macaca fascicularis* [Gathorne-Hardy (Lord Medway), 1978]. Monkeys occasionally occur on Seram, as pets in major settlements, where they may have been introduced by rajas (as in other parts of Indonesia) or travellers and sailors, but not in Nuaulu villages. All available historical and archaeological data suggest that they have never been endemic to the island. They feature in some Nuaulu myths and stories.

The monkeys with which the Nuaulu are most familiar are possibly the long-tailed macaque (*Macaca irus*) and the Celebes macaque (*Macaca maurus*). *Cynopithecus niger,* the crested Celebes macaque, is found in Bacan to the north, where it was probably introduced by human agency. This monkey may also be known to the Nuaulu.

2.2.19 gaja

This term was applied to descriptions of the asiatic elephant (*Elephas maximus*), known only to the Nuaulu through its reputation, verbal descriptions and graphic representations. From AM 'gaja'.

2.2.20 rompa

Rosemary Bolton has recorded this term (from the 'domba') for sheep. There are no sheep on Seram and this almost certainly recent lexical introduction (perhaps since 1970) may owe something to Christian biblical allusion.

2.2.21 maisan

Rosemary Bolton [personal comm.] reports that two different Nuaulu informants applied the term to 'lion' and 'tiger'. In my experience Nuaulu refer to these unencountered creatures using AM 'singa' and 'harimau' respectively, though it is possible that **maisan** is used as a general-purpose label for various strange and exotic beasts with which they are not familiar. On one occasion **maisan** was applied to descriptions of an elephant, otherwise **gaja**. The term **maisan** appears to be from 'macan', in Javanese and other Indonesian languages the word for tiger.

2.3 More inclusive categories for mammals

There is no general term remotely corresponding to English MAMMAL, and Cecil Brown's suggestion [Brown, 1984: 227] that **peni** might be a pseudo- or proto-mammal term is misleading. Some aspects of this complex ritually salient category are treated extensively in chapter 4 of *The Cultural Relations of Classification*. There are only two terms which indicate clearly the existence of larger categories dividing up the phylogenetic space for MAMMAL. The first is **marane**. The second is **mnahane,** which in its most inclusive sense is used for all murids. There are also certain *ad hoc* juxtaposed uninomials which bring together various large mammalian quadrupeds with a perceived resemblance: **sapi-korobou, hahu-maianane** and **lau-kuha.**

Notes to Chapter 2

1   The taxonomy of the family Phalangeridae has recently [Macdonald et al, n.d.] been revised, such that *Phalanger maculatus* becomes *Spilocuscus maculatus*. For the sake of consistency with earlier publications, and because this revision only came to my attention fairly late on in the preparation of this monograph, I have retained the older nomenclature here.

2   In the field specimens were examined both alive and dead, in the course of hunting expeditions, prior to preparation for eating in the village and in the ritual context of male initiation ceremonies. It is difficult to estimate accurately how many separate specimens were examined during the course of the first eighteen months' fieldwork, but judging from records kept for household consumption of cuscus it must have been in the region of 65 animals. For a period of four months I kept a male *Phalanger orientalis* in captivity, which gave me an opportunity to

observe some of its habits at close quarters and served as a useful aid in discussing cuscus natural history with informants. Unfortunately, due to bulk and difficulties in transportation, only one spirit specimen was subsequently examined by the Natural History Museum in London, a young male *P. orientalis orientalis.* Four skulls of specimens identified by the Nuaulu were examined following the 1975 field trip.

3   **Mara makioi** is prohibited as food for a **kapitane**, although I am unsure whether this restriction applies to all clans.

4   **Mara porune** was contrasted by some informants with **mara onate,** and this usage appears to apply to all four terminal adult categories. Although **onate** can accurately be glossed as 'large, big', **porune** cannot be glossed simply as 'little, small' (**ikine**), despite the fact that in contrast to **onate** this must be one of its referents. It is a genus-specific term, being used for no other animal, and is therefore similar to such English folk usages as 'cygnet' or 'lamb'. However, in certain contexts the term was used to differentiate an individual from any one of the four adult forms. The reason for this appears to be that the characteristics (which are basically in terms of size and fur coloration) are not always visibly apparent in younger members of the genus.

5   A detailed description of hunting practices has been published elsewhere [Ellen, 1972]. In that account I discuss some common cuscus habitats. I would now wish to add to the places mentioned *Ficus, Canarium* and *Durio* trees, whose leaves they eat. Further details on the dietary significance of cuscus and other animals are provided in chapter 6.3 of *The Cultural Relations of Classification.*

6   This statement supersedes an earlier erroneous claim that the Nuaulu terms **lau** and **kuha** refer to *Rhynchomeles* [Ellen, 1972: 225; Ellen, 1975a: 203].

CHAPTER THREE

BATS

3.1 The bat fauna of south central Seram

In a region where the vertebrate fauna is notoriously depauperate, and mammal fauna more depauperate than that of most classes, bats (with birds) represent something of an exception. This is largely due to their diffusion not being subject to the same geographical constraints as other groups.

Prior to 1969, 19 certain species of *CHIROPTERA* were known from Seram. There were also three probables: *Rousettus amplexicaudatus* (the rousette or dog bat), *Nyctimene minutus* (a tube nosed bat) and *Rhinolophus keyensis* (a horshoe bat). Ten of these were observed and collected during fieldwork in the Nuaulu area between 1970 and 1975, including a specimen of *Rhinolophus keyensis* - probably of the subspecies *truncatus*. The collection also included specimens of *Emballonura nigrescens nigrescens*, previously only recorded in the Central Moluccas from Ambon and Buru. The specimens of *Macroglossus minimus* from Seram have been attributed to the subspecies *lagochilus*.

The difference between the number of species collected and those known zoologically from the entire island may be accounted for by relative geographical distribution, particularly differences between the highlands and the coast. A checklist of bats reported from south central Seram is presented in table 4. Species identifications compared with Nuaulu categories applied to actual specimens are set out in table 5.

3.2 Nuaulu categories applied to bats

3.2.1 nota sapane

Sapane is also the term elicited for the tree *Trema* sp. prob. *cannabina,* widely associated by the Nuaulu with secondary regrowth in swiddens. All specimens of *Pteropus melanopogon melanopogon* were given this name. This large tree-living fruit bat is extremely common. On the basis of evidence from one informant (Sorita), the term is also employed to refer to fruit bats in general, of which two kinds are distinguished: **nota sapane** in its specific sense (i.e. *P. melanopogon melanopogon*) and **nota maka paluwe.**

TABLE 4 Checklist of bats (CHIROPTERA) recorded in the Nuaulu region of south central Seram.

Species	Ecological Zones 1 2 3 4				Nuaulu glosses

MEGACHIROPTERA

Pteropodidae
Pteropodinae - flying foxes

Species	1	2	3	4	Nuaulu glosses
Pteropus melanopogon melanopogon bare-backed fruit bats	?	+	+	?	**nota sapane**
Dobsonia viridis viridis	?	+	+	?	
Dobsonia moluccensis moluccensis	?	+	+	+	**nota hatu nohue** **nota maka paluwe**

Macroglossinae - long tongued fruit bats

Species	1	2	3	4	Nuaulu glosses
Macroglossus minimus	+	+	+	-	**nota muni munte,** **nota kiniku putie**
Syconycteris crassa major	?	+	+	-	**nota muni munte,** **nota kiniku putie**

Nyctimeninae - tube nosed bats

Species	1	2	3	4	Nuaulu glosses
Nyctimene sp.	?	+	+	-	**nota muni munte,** **nota kiniku putie**

MICROCHIROPTERA

Emballonuridae
Emballonurinae - sheath-tailed bats

Species	1	2	3	4	Nuaulu glosses
Emballonura nigrescens nigrescens	?	+	+	-	**notane**
Emballonura raffrayana stresemanni	?	+	+	-	**nota mana waute,** **nota muni munte,** **nota kiniku putie,** **nota maka paluwe**

Rhinolophidae

Rhinolophinae - horseshoe bats

Rhinolophus keyensis (?) truncatus   ?   +   +   -

Hipposiderinae - leaf nosed bats

Hipposideros ater saevus   ?   +   +   -   **nota muni munte,
nota kiniku putie,
nota hatu nohue**

Notes: : *Key*: Zone 1 = above 1000 metres, principally montane rain forest; Zone 2 = tropical lowland rain forest; Zone 3 = secondary forest, garden and village areas; Zone 4 = freshwater and swamp forest.

3.2.2 nota maka paluwe (baluwe, paloi)

As large as **nota sapane,** but white band with rufous tinge round back of neck to head; brown head, blackish body. This term was consistently applied to specimens of *Dobsonia moluccensis moluccensis.* Large numbers of this bat are visible in the extensive sago swamp forests towards the mouth the river Ruatan. The Nuaulu associate large flocks of this species on the wing with heavy rain. In the evening clouds of them can sometimes be seen swarming from these swamplands to feed in the fruit-rich forests to the north.

3.2.3 nota hata nohue

Hata nohue, meaning 'cavernous rock outcrop', 'cave', indicates the habitat of this bat. It is found particularly in the extensive limestone caverns of the Lahati area. All informants regarded **nota hatu nohue** as smaller than **nota sapane,** and indeed the overwhelming number of specimens described in this way were *Emballonura raffrayana stresemanni.* However, nine specimens of *Dobsonia moluccensis moluccensis,* three of *Dobsonia viridis viridis* and two of *Hipposideros ater* were also given this term. In view of this diversity of content, and since other terms were also applied to all of these species, I am inclined to treat **nota hatu nohue** as a generic term for all cave-dwelling bats. On the other hand, it is clear that in some contexts it may contrast with terms applied to the same species. For example, informants might remark, with reference to a specimen of *Emballonura raffrayana stresemanni,* that it was not **nota muni munte** but **nota hatu nohue.** Thus, although it may apply to a number of species which are also described in other ways, I feel that it must be considered to indicate a valid terminal category.

Table 5   Species identifications compared with Nuaulu categories applied to 81 bat specimens

Species	nota sapane	nota hatu nohue	nota maka paluwe	nota suite	nota wanu anae	nota mana waute	nota muni munte	nota kiniku putie	notane	Unknown	Informant responses	No responses available	Number of specimens
Pteropus melanopogon melanopogon	3	-	-	-	-	-	-	-	-	-	3	-	1
Dobsonia viridis viridis	-	3	-	-	-	-	-	-	-	-	3	-	1
Dobsonia moluccensis moluccensis	-	9	4	-	-	-	-	-	-	-	13	4	9
Macroglossus minimus	-	-	-	5	4	-	5	16	-	1	31	-	14
Synconycteris crassa major	-	-	-	-	-	-	6	6	-	3	15	-	3
Nyctimene sp.	-	-	-	-	-	-	1	3	-	-	4	-	1
Emballonura nigrescens nigrescens	-	-	-	-	-	-	-	-	9	-	9	-	9
Emballonura raffrayana stresemanni	-	108	-	-	-	7	54	54	-	-	223	-	34
Rhinolophus keyensis (?) truncatus	-	-	-	-	-	-	-	-	-	-	-	1	1
Hipposideros ater	-	12	-	-	-	-	36	16	-	-	64	-	8
Totals	3	132	4	5	4	7	102	95	9	4	365	5	81

Nuaulu categories

3.2.4 nota suite

This term, perhaps an allusion to the bird of the same name (Chapter 4.2.51), was consistently applied to specimens of *Macroglossus minimus,* to which three other terms were also applied.

3.2.5 nota wanu anae

Wanu anae seems to mean 'younger brother's son' or perhaps more generally 'brother's child' (**wanim panui**) or 'sister's child' (**wanim hotam**); **waniku** = 'my younger brother'. **Wanunui** is the title of an **ahinae,** a form of chant which accompanies the **kahuae** circle dance. This term was consistently applied to specimens of *Macroglossus minimus,* to which three other terms were also applied.

3.2.6 nota mana waute

The term for this bat is possibly an allusion to a beard (**manapesi**) as a distinguishing feature. The term was consistently applied to *Emballonura raffrayana stresemanni,* to which three other terms were also applied. The possible reference to a beard is odd. *E. r. stresemanni* does not possess one, although *Tophozous melanopogon* does. This latter species is known from Savu and Sumbawa, but not Seram.

3.2.7 nota muni munte

This term for a small cave bat was applied to *Macroglossus minimus, Syconycteris crassa major, Nyctimene* sp., *Emballonura raffrayana stresemanni* and *Hipposideros ater.* One informant was emphatic in its identification with *Nyctimene* (emphasizing its tube nose) (Sorita Matoke). Konane Nepane-tomoien described it as having long ears with a 'horseshoe' shape over the nose, although Sorita stressed that the male (identified as *Nyctimene*) did not have long ears, although the female did. In this he agreed with the opinion of Sekanima Nepane-tomoien.

3.2.8 nota kiniku putie

Literally, the 'white sago palm bark bat'. This was said by Naupati Matoke to be common, occurring singly rather than in groups. It has a white head, is of medium size and forest-dwelling. According to Naupati it is difficult to catch. The term was applied to specimens of *Macroglossus minimus, Syconycteris crassa major, Nyctimene* sp. and *Emballonura raffrayana stresemanni.*

3.2.9 kupapa

Term elicited by Bolton; unrecorded by Ellen. Possibly a synonym motivated by gender or clan linked prohibitions.

3.5 Social uses of bats

There are no totemic or other prohibitions on eating bat meat which I have been able to discover. In the course of the four-month dietary survey conducted in 1970 approximately 75 grams of bat meat were consumed per adult head. Although this was only 0.5 percent of the total weight of animal protein consumed, it rated seventh in importance in terms of all food from terrestrial sources. All bats will be eaten if available; even the small specimens caught in mist nets by Menzies and myself were accepted. However, on the whole, only larger bats will be actively hunted. *Dobsonia moluccensis moluccensis* (**nota maka paluwe**) are a main source of food for men working sago palms in the swamp forests of the Ruatan-Nua confluence, where they are shot with a bow-and-arrow. *Pteropus melanopogon melanopogon* (**nota sapane**) is hunted in the same way in forest nearer the village, where it may occasionally damage cultivated trees.

Dobsonia, and perhaps also other species, are hunted in lofty limestone caves in the Lahati and Lihuru areas, and possibly elsewhere. The bats are located and disturbed using noise and firebrands of dried coconut fronds which have to be brought to the caves from garden areas. Makeshift ladders are erected and strong branches used to reach otherwise inaccessible recesses in the cave wall and roof. Throwing-sticks and bow-and-arrows are used to immobilize the bats, although arrows quickly become blunted when used in this way. Cave floors are littered with the charred remains of torches and sticks, and are also covered in deep deposits of accumulated bat dung in which are embedded the shells of thousands of terrestrial molluscs.

Apart from food, the only use recorded for bats is in the making of smoking-pipes (**suparko**) from the hollow long bones of the larger fruit bats [Ellen 1993: plate 1.6f].

3.6 Variation in the identification and classification of bats

The only species unambiguously and consistently linked to Nuaulu categories are *Pteropus melanopogon* (**nota sapane**) and *Dobsonia viridis* (**nota hatu nohue**). One species, *Emballonura nigrescens* was identified only as **notane.** *Dobsonia moluccensis moluccensis* was identified as both **nota hatu nohue** and **nota maka paluwe,** and it seems that **nota hatu nohue** is used generically to cover a number of species found in caves (see section 3.2.3, above). In view of this I assume the conflicting identifications to be based on habitat associations, and suspect that specimens of *D. viridis viridis* not found in caves would be identified as **nota maka paluwe.**

Of the remainder, there are five terms and six species. Of these, **nota suite, nota wanu anae** and **nota waute** appear to have a one-to-one correspondence with phylogenetic species, though Rosemary Bolton has suggested that the terms **nota wanu anae, nota waute** and **nota kiniku putie** might sometimes be used synonymously.

The two terms employed most frequently are **nota muni munte** and **nota kiniku putie.** This suggests that these are -in general- the most widely known and used terms. This is supported by their more frequent appearance in lists of the total numbers of terms elicited from individuals. Given that they seem to be employed virtually interchangeably for at least five species, neither seems to be indissolubly linked with definite criteria of distinction. However, since they are frequently contrasted, they cannot be treated as synonyms. The species and sub-species most frequently identified as either **nota muni munte** or **nota kiniku putie** are *Emballonura raffrayana stresemanni* and *Hipposideros ater,* and on crude statistical grounds it is possible that **nota kiniku putie** is focally the first and **nota muni munte** focally the second.

No terminal term was elicited for *Rhinolophus keyensis (?) truncatus.*

A taxonomic representation seems particularly unsuitable for Nuaulu classification of bats. The main contrasts are between fruit bats and residual categories, and between cave bats and residual categories; although fruit bats and cave bats may overlap inasmuch as specimens of *Dobsonia* (identified as **nota hatu nohue**) were obtained from caves in the Lahati area. The remaining bat species are generally assigned either to the category **nota kiniku putie** or **nota muni munte.** As we have seen, there is some evidence for the focality of these categories, but while some individuals may assert firm identities between specimens and named categories, some of the terms appear to be used as relatively 'loose labels'.

CHAPTER FOUR

BIRDS

4.1 The avifauna of south central Seram

Van Bemmel [Bemmel, 1948; Bemmel and Voous, 1953] lists 161 certain bird species for the mainland of Seram, plus two probables. By 1989 [Bowler and Taylor, 1989] this figure had been pushed up to 195. Of these, five were collected as spirit specimens during fieldwork in the Nuaulu area between 1970 and 1975, 33 were observed and identified to species level, five to generic level, five to family level and five additional distinct types to order level. These included two introduced species not listed in the ornithological sources for Seram: *Anas boscas* (domesticated duck), and *Gallus gallus* (domesticated fowl). Omitting these two species gives 46 species observed, compared to the 195 reported by others.

The difference between the number of species observed during fieldwork and that reported for the entire island may be accounted for by relative geographical distribution, particularly differences between the highlands and the coast. For example, certain species listed in the literature are given specifically for mountainous areas (e.g. *Zosterops montanus seranensis*). I suspect that many birds not observed during fieldwork are present in the Nuaulu region, both permanently and seasonally as migrants; others appearing in checklists are rare migrants or stragglers. According to Bowler and Taylor, 71 species are true migrants. The migratory character of many bird species presents a special factor not present for other animal groups when we come to explain the ratio of Nuaulu to phylogenetic categories, and in assessing the significance of variable degrees of knowledgeability of informants.

A checklist of birds observed in south central Seram, or whose existence is strongly suspected, is presented in table 6.

4.2 Nuaulu categories applied to birds

The inventory which follows is based entirely on terms elicited during the course of my own fieldwork. Stresemann [1914] provides some Nuaulu terms, though they are not always applied in ways which are consistent with my own observations. I have indicated such terms and determinations where I have felt this to be instructive. Stresemann lists just two terms which I have not been able to locate in contemporary Nuaulu: **manu asan** for *Merops ornatus,* the migrant Rainbow bee-eater [*ibid.,* 109]; and **sisa une** for *Criniger affinis.*

4.2.1 asuwan, rui-rui

This term always and unambiguously refers to the double-wattled cassowary, *Casuarius casuarius*. Immature cassowaries are called **rui-rui,** and this term (apparently the older of the two) is sometimes used synonymously for the adult. **Asuwan** appears to be replacing **rui-rui.** Certainly between 1970 and 1990 it was that most prominent in ritual contexts, with respect to the more inclusive category **peni,** and in certain linguistic idioms such as the expletive **asuwani anae!** [Ellen 1993, chapters 4.4 and 6.4].

Mature birds of both sexes can be up to 183 cm in height (ranging between 122 and 183 cm), with a blade-like horny casque on the crown

(**pipane**), often turned to the left. The wattle tends to vary geographically in shape; the colour of the bare neck and head is mostly blue and red or pink, sometimes with some yellow; young birds have a brown plumage, with much pink and yellow in the skin of the neck; younger birds have the neck feathered; newly hatched chicks are brown, streaked black [Rand and Gilliard, 1967: 21-3]. The taxonomic and phylogenetic status of the Seramese form is unclear. It has been suggested [Rand and Gilliard, 1967: 23; White, 1975: 165-10] that it was introduced into the island from New Guinea at least as long ago as the twelfth century. Sachse, 1907: 54 suggests that its distribution is mainly in east Seram, but it is abundant in the Nuaulu area. It is usually solitary, though mature females may have up to five dependent chicks at a time.

Cassowary appears to rate about sixth in importance as a source of animal protein, contributing three percent of that consumed per head per day over a four month period and being eaten on only eight occasions out of a sample of 507 meals. The plumes of the tail region of the cassowary are used for head-dresses (**orane**), worn by clan chiefs and **kapitane** [Ellen 1993: frontispiece and plate 1.6e], and used to decorate armbands [*ibid*, plate 1.6a], when they are described as **sinsin asuwani hunue,** 'cassowary feather **sinsinte**'. **Sinsinte** is the cordyline *Codiaeum variegatum,* used ordinarily for trailing arm decorations and gives its name to such decorations as a generic type. The armbands themselves (**nonie**) are made from the split thick quills of the vestigial wings, rattan and bark [*ibid*, plate 1.6j]. Cassowary claws (BM As 1.40a and 40b) and eggs (BM As 1.251) are also used for decorative purposes, and occasionally as lime containers. Leg bone is used for bradawls or **kahate,** employed mainly in the manufacture of containers from the leaf-stalk of the sago palm [*ibid*, plate 1.6c].

The classificatory and ritual significance of the cassowary has been discussed at some length elsewhere [Ellen, 1975a], and is considered in *The*

TABLE 6 Checklist of birds recorded in and around the Nuaulu region of south central Seram.

Species	Ecological zones					Nuaulu gloss
	1	2	3	4	5	
CASUARIIFORMES						
Casuariidae - cassowaries						
Casuarius casuarius	+	+	+	-	-	**asuwan**
PELECANIFORMES	-	-	-	-	+	**nusi tanane**
Sulidae - gannets and boobies						
Sula sula rubripes						
S. leucogaster plotus						
Phalacrocoracidae - cormorants						
Phalacrocorax melanoleucos						
P. sulcirostris						
Fregatidae - frigate birds						
Fregata ariel						
F. minor						
CICONIIFORMES						
Ardeidae - hersons and allies	-	-	-	+	-	**sote**
Dupetor flavicollis						
prob. *Bubulcus ibis*						
prob. *Butorides striatus*						
prob. four species of *Egretta*						
prob. *E. garzetta*						
prob. *E. intermedia*						
prob. *E. sacra*						
prob. *Ixobrychus sinensis*						
Nycticorax caledonicus						
Phalaropididae						
Phalaropus lobatus						
Scolopacidae						
Actitis hypoleucos						
and others						
Threskiornithidae						
Threskiornis moluccus,	-	-	-	+	-	**moinapu**
white ibis						
uncertain	-	-	-	+	-	**samane**
ANSERIFORMES						
Anatidae - ducks, geese etc						

Anas boscas, domesticated duck	-	-	+	-	-	pepeko
FALCONIFORMES						muinu
Accipitridae - harriers, hawks and eagles	+	+	+	-	-	muinu toa
Accipiter cirrhocephalus ceramensis, *Accipiter novaehollandiae* and possibly others						
Pandionidae - ospreys *Pandion haliaetus*	-	-	-	+	+	nusi takanasi
Falconidae - falcons *Falco moluccensis* *F. peregrinus*	-	+	+	-	-	muinu puane
GALLIFORMES						
Megapodiidae - megapodes *Megapodius reinwardt* orange-footed scrubfowl	-	+	+	-	-	kinosane
M. wallacei Moluccas scrub hen	-	+	+	-	-	muma
Phasianidae - quails and pheasants *Coturnix chinensis lineata*, Chinese quail	-	-	+	-	-	kowe marae (♂), kowe metene (♀)
Gallus gallus, domesticated fowl	-	-	+	-	-	man tulalakane (♂), man winai (♀)
GRUIFORMES - rails and related forms	-	+	-	+	-	sote nakone
Amaurornis olivaceus *Gallinula tenebrosa* *Poliolimnas cinereus* *Porphyrio porphyrio* *Porzana pusilla* *Rallus philippensis* prob. *Rallina? fasciata*						
CHARADRIIFORMES Charadriidae - plovers	-	-	-	+	+	hiko-hiko
Charadrius leschenaultii *C. mongolus* poss. *Squatarola dominicus*						

Laridae - gulls and terns
Chlidonias hybridus,

marsh tern	-	-	-	+	-	nusi kiene
Sterna bergii	-	-	-	+	-	nusi onate
other terns	-	-	-	-	+	nusi puane, nusi seane

at least three species of *Sterna*

COLUMBIFORMES
Columbidae - pigeons and doves

Macropygia amboinensis, Amboina cuckoo-dove	-	+	+	-	-	puane

prob. *Gymnophaps mada*,
long-tailed mountain pigeon
Ducula bicolor,

white nutmeg imperial pigeon	-	+	+	-	-	hutiene, manu pake

Reinwardtoena reinwardtii,
great cuckoo dove

PSITTACIFORMES
Psittacidae - lories, parrots and cockatoos
Lorius domicella,

black-capped or purple-naped lory	-	+	+	-	-	kihoke

also *Micropsitta bruijnii*
Cacatua moluccensis,

Moluccan Cockatoo	-	+	+	-	-	nakatua (putie)

Cacatua galerita
Eclectus roratus,

red-sided eclectus parrot	-	+	+	-	-	wekae msinae

Tanygnathus megalorhynchos,

Moluccan or island parrot	-	+	+	-	-	konane

various Lories, including
Eos semilarvata,

blue-eared lory	-	+	+	-	-	wekae marae

E. squamata
poss. *Charmosyna placentis*
E. bornea,

and possibly other related species	-	+	+	-	-	kunini

Alisterus amboinensis,

Amboinan king parrot	-	+	-	-	-	siseite

Geoffroyus geoffroyi,

red-cheeked parrot	-	+	+	-	-	tuie

CUCULIFORMES
Cuculidae - cuckoos
  Cuculus saturatus,
  oriental cuckoo   -   -   +   -   -   **kukue**
  Caculus variolosus,
  grey-breasted brush cuckoo
  two species of *halcites*
  Eudynamys scolopacea orientalis,
  Indian koel
  Scythrops novaehollandiae,
  channel-bill cuckoo
  Centropus bengalensis medius,   -   -   +   -   -   **suanane**
  lesser coucal

STRIGIFORMES
Strigidae - owls
  Otus magicus   -   +   +   -   -   **tuku-tuku**
  Ninox squamipila   -   +   +   -   -   **sakoa**

CAPRIMULGIFORMES
Caprimulgidae - nightjars
  Caprimulgus macrurus,
  large-tailed nightjar   -   +   +   -   -   **kuna-kuna**
  uncertain   -   +   +   -   -   **man totane**

APODIFORMES
Hemiprocnidae
  Hemiprocne mystacea confirmata

Apodidae - swifts
  Collocalia esculenta esculenta,   -   +   +   -   -   **sioi, kina nahane,**
  glossy swiftlet   **neune anae**
  Collocalia spodiopygia ceramensis

CORACIIFORMES
Alcedinidae - kingfishers
  Halcyon sancta,
  sacred kingfisher   -   -   +   +   -   **tialapeti**
  uncertain, perhaps
  Halcyon chloris   -   +   +   -   -   **mui numte**
  Halcyon lazuli
  Ceyx lepidus,
  dwarf kingfisher   -   +   +   +   -   **man tuamane**
  Tanysiptera galatea,
  common paradise kingfisher   -   -   -   +   -   **saneane**
  also *Ceyx*
  Alcedo atthis hispidoides

	1	2	3	4	5	
Bucerotidae - hornbills *Rhyticeros plicatus*, Papuan hornbill	-	+	-	-	-	**sopite**

PASSERIFORMES

	1	2	3	4	5	
Pachycephalidae - shrike flycatchers and whistlers *Pachycephala pectoralis*, golden whistler	-	+	+	-	-	**soihihi**
Sturnidae - starlings incl. *Basilornis corythaix*, long crested myna *Aplonis metallica* *A. mysolensis forsteni*						
Dicruridae - drongos *Dicrurus bracteatus*	-	+	+	-	-	**tika poho-poho**
Corvidae - crows	-	-	+	+	-	**tika poho-poho, opor**
Corvus enca violaceus						
Nectariniidae - sunbirds *Nectarinia aspasia*						
prob. *N. jugularis clemetiae*	-	+	+	-	-	**tinnanae**
Meliphagidae - honey eaters *Philemon subcorniculatus*, Seram friar bird *Lichmera monticola*, Seram honey eater	-	-	+	-	-	**kinoke**
uncertain perching birds						
	-	-	+	-	-	**katenane suite sopate**
	-	+	+	-	-	**hiru**
uncertain birds of all orders						
						manu kasa neane kura-kura

Key: Zone 1 = above 1000 meters, principally montane rain forest; zone 2 = tropical lowland rain forest; zone 3 = secondary forest, garden and village areas; zone 4 = freshwater and swamp forest; zone 5 = birds of the sea and littoral.

Cultural Relations of Classification in relation to anomaly and salience (chapter 6.7) and in relation to the functioning of the non-basic categories **manue** and **peni** (chapter 4.4). It is considered further here in section 4.3.3.

4.2.2 nusi tanane

Nusi also means 'citrus fruit', though the two different usages are almost certainly homonymous only and not semantically cognate. The meaning of **tanane** is unclear: **tana** is sometimes used for 'earth' or 'ground', though the connection with the habits of these birds is not apparent. Occasionally the same birds may be referred to as **manu tanane.**

The term is applied fairly indiscriminately to various kinds of seabirds of the order PELECANIFORMES, including boobies (*Sula sula, S. leuco-gaster*), cormorants (*Phalacrocorax melanoleucos, P. sulcirostris*) and frigate-birds (*Fregata ariel, F. minor*). I have been unable to identify any particular focal or type species.

4.2.3 nusi takanasi

A **takanasi** is a type of large basket used for storage purposes and may refer to the distinctive nest of this bird. The term appears to be applied very specifically to the osprey, *Pandion haliaetus,* and is glossed in AM as 'burung elang eking'. It is said to live mainly from sea fish, but is known to have attacked both cuscus and dogs.

The myth of origin of this bird, which is said to come from Hatu Takanasi (i.e. **takanasi** stone) on the river Nua, is enshrined in a stanza of the tug-of-war song, the **kepata ararirane.**

4.2.4 (nusi) kiene

Kiene is an obvious onomatope, in this case for terns. The category includes *Chlidonias hybridus,* the marsh tern, but the precise range of content is unclear.

4.2.5 nusi onate

Probably a large tern (**onate** = 'large'), possibly *Sterna bergii,* the greater crested tern.

4.2.6 nusi puane

Puane alludes to another bird category (4.2.22) which includes the Amboina cuckoo-dove, *Macropygia amboinensis.* **Nusi puane** is said to be the equivalent to 'burung camar' in AM and 'mantel-meeuw' in Dutch, which Iskandar, 1970: 159 gives as *Thalasseus* (i.e. *Sterna bengalensis*). The Nuaulu apply the term to all or some of the following Seramese terns: *Sterna anaethetus, S. hirundo, S. sumatrana.* I assume that the term **puane** relates to the superficial resemblance between terns and doves.

4.2.7 nusi seane

Seane means 'wild', and is only used in connection with birds. As with **nusi puane,** this term appears to refer to a tern, and is also glossed with AM 'burung camar'; presumably one or more of the species listed under 4.2.6.

4.2.8 (manu) nusi

Judging from the data presented in sections 4.2.2 through to 4.2.7, the term **nusi** refers to most of the larger birds of the sea coast plus some morphologically related birds of the rivers and inland waters, including all resident and itinerant *PELECANIFORMES* known to the Nuaulu, one family of *FALCONIFORMES* (the osprey) and one family of *CHARADRIIFORMES* (gulls and terns). All these birds are fish-eaters, although they will occasionally eat small land animals. At least one informant applied the term to the broad-billed roller, *Eurystomus orientalis pacificus.* Rosemary Bolton reports the AM gloss 'garuda rajawali' applied to this category.

4.2.9 sote

The term is used polysemously. In its narrow sense it refers to a category which includes herons and their allies, although Stresemann, 1914: 67, gives it as the gloss for *Dupetor flavicollis.* One informant described the term as referring to large forest-dwelling birds, similar to chickens, with habitats not far from human settlement and eating fruits such as those of *Canarium indicum,* the kenari almond. The reference to eating habits is a little curious, although it is known to have a very varied diet. It seems that the category includes several herons, egrets and bitterns known from Seram, namely *Egretta* spp., *Butorides striatus, Bubulcus ibis, Ncyticorax caledonicus* and possibly the migrant *Ixobrychus sinensis.* In its widest sense, **sote** indicates a category which also includes rails.

4.2.10 sote nakone

This term refers to a category which includes rails and related forms: *Amaurornis olivaceus, Gallinula tenebrosa, Poliolimnas cinereus, Porphyrio porphyrio, Rallus philippensis, Porzana pusilla* and possibly *Rallina fasciata.* The similarities bringing herons (4.2.9) and rails together terminologically, and which are stressed by the Nuaulu, are long legs compared with body, curvaceous necks, long beaks and a fish diet.

4.2.11 moinapu (mui napu)

Moi is a prefix of respect (third person singular, inalienable possession), related to the relationship term **momo-** (MM,MF,FF,FM, DD, DS, SS, SD); **muine** means 'origin'. **Napu** is a gourd, *Lagenaria leucantha*. This term labels a category focussed on the white ibis *Threskiornis moluccus*, but extending to a large group of related migrants (sandpipers and their allies): certainly *Phalaropus lobatus* and probably *Actitis hypoleucos*. However, the gourd allusion is not employed when referring to the morphologically very similar herons and their allies. **Napu** may refer to the curvaceous gourd-like neck of birds assigned to this category, or to the shape of the bill.

4.2.12 (man) samane

Described by one informant as a white, forest-dwelling, riverine fish-eater. Undetermined stork, heron or ibis.

4.2.13 pepeko

Cognate with AM 'bebek', meaning 'duck'. Refers to all domesticated waterfowl, which are almost always *Anas boscas*. May also refer to *Anas moschata*, the muscovy duck, which is sometimes found in Indonesia. The Nuaulu seldom keep ducks, although many of the peoples surrounding them do. There are a few in some hamlets, mainly to produce eggs and young adults for the market. There appear to be no restrictions on consumption. Four species of wild duck are reported for Seram: *Anas querquedula, Nettapus pulchellus, Tadorna radjah* and *Denorocygno guttata*. None of these were observed, the first of which is a migrant and the second a possible winter visitor. There appear to be no Nuaulu categories other than **pepeko** to cope with them and this term suggests a borrowing quite specifically to label domesticated types.

4.2.14 muinu puane

As pointed out at 4.2.6, **puane** employed uninominally refers to a category which includes the Amboina cuckoo-dove, *Macropygia amboinensis*. I suspect that both **nusi puane** and **muinu puane** are so-called because of their superficial resemblance to the **puane**. Nuaulu assign to this category the genus *Falco*, of which four species are known from Seram: *F. cenchroides, F. longivennis, F. moluccenis* and *F. peregrinus*. As the second and fourth of these are migrants, and the first a straggler from Australia, the focus would appear to be *F. moluccensis*.

4.2.15 muinu toa

Toa- is a relationship term for those persons five generations ascendant or descendant from ego (e.g. FFFFF, MFFFF, SSSSS, DSSSS). Labels a category which includes various harriers, hawks and perhaps eagles: focally *Accipiter cirrhocephalus, A. novaehollandiae;* but may also be extended to *A. meyerianus, Aviceda subcristata, Haliaeetus leucogaster, Haliastur indus* and *Ictinaetus malayensis.*

4.2.16 muinu, moi

See 4.2.11. A generic term labelling the *FALCONIFORMES* listed in 4.2.14 and 4.2.15. In describing species labelled **muinu** informants repeatedly emphasised their carnivorous habits, saying that they occasionally preyed on domesticated chickens, particularly young ones, and were known to also eat cuscus (2.2.1). They are considered vermin. Another feature emphasised, excluding size, shape and colour, was the ability of **muinu** to hover for long periods. Despite the lexical similarity with **muinapu,** the terms do not appear to be related (see also **mui numte,** 4.2.40).

4.2.17 kinosane, manu asu

Manu asu (lit. 'dog bird') is almost certainly a true synonym here. The orange-footed scrubfowl, *Megapodius reinwardt.* The Nuaulu compare it with domesticated chickens and **sote** (4.2.10), and describe it as fruit-eating and living near fresh water. Its flesh and eggs are occasionally eaten.

4.2.18 muma

The Moluccas scrub hen, *Megapodius wallacei,* known in AM as 'maileu'. Both the flesh and large eggs are eaten, and the latter are particularly sought after by the Nuaulu. The flesh is said to be good, and much more tender than that of the cassowary.

M. wallacei lays its eggs in the warm sands of certain headlands in the Nuaulu area, above the high-water mark, each nest sometimes containing as many as 20 eggs. Small nests are excavated by the female which are then covered over. In collecting eggs, Nuaulu stress the importance of being able to recognise an excavation of the previous night, otherwise growth of the fertilised eggs will be too advanced. Eggs are therefore collected very early in the morning. Eggs are not laid in the rainy season. Adult *Megapodius* are difficult to locate, as they are not only nocturnal, but also retreat to the forest before daylight. They can, however, sometimes be caught using a special trap (**hai unai**).

4.2.19 (manu) mkowe: mko marae, mko metene

The second two terms are complementary for the male and female respectively of a sexually dimorphic species. **Mkowe** is possibly a contracted form of **mara kokowe,** male of the cuscus species *Phalanger orientalis orientalis* (chapter 2.2.1). **Marae** refers to the blue-grey markings on the underparts of the male and **metene** to the black bars present on the breast and sides of the female. The AM qualifiers 'abu' and 'hitam' respectively are sometimes substituted. This is a subspecies of the Chinese quail, *Coturnix chinensis lineata.*

4.2.20 man tulalakane, manu inae

The prefix **man** comes from the root **manu,** which also refers to 'birds' in general. **Manu** and **manue** are in free variation with **man,** at least with respect to **man tulalakane.** As with **mkowe,** the qualifiers terminologically distinguish between males (**tulalakane**) and females (**inae**), the former being an onomatope. The terms gloss as the cock and hen respectively of the domesticated chicken, *Gallus gallus* [1]. **Manu inae** ('mother bird') is occasionally used as an inclusive term for both sex-types, although only in contexts in which a polemical or illustrative point is being made concerning either the mythical origin of all birds or the folk-etymology of the category **manue.** Rosemary Bolton also reports the (collective) terms **sihunane** and **muna nea.**

Nowadays a large number of Nuaulu households keep chickens which feed on domestic scraps. The coarse residue of sago flour processing may sometimes be made available to them, but Nuaulu chicken husbandry is otherwise minimal. Special receptacles for hens to lay their eggs in are attached to the sides of houses, usually made out of an old conical fish trap (**kananesa**). Chicken eggs are rarely eaten by the Nuaulu themselves, and chicken flesh even less. The eating of chickens is ritually restricted for the head of the clan Matoke. Chickens are primarily reared for the sale of live birds and eggs to non-Nuaulu.

4.2.21 hiko-hiko (hiko)

The term is probably onomatopoeic, a judgement supported by the preferred reduplicated form, **hiko-hiko.** Applied to small light-coloured migrant plovers of the genus *Charadrius,* commonly seen on beaches in the appropriate season. They occur singly or in small groups, usually making short sorties over water and returning to the beach. It is possible that the term is also extended to other *CHARADRIFORMES,* for example *Squatarola dominicus.* Stresemann [1914: 128] uses the term **tiko-tiko** (which must surely be the same) to refer to specimens of the flycatcher *Monarcha trivigatus.*

4.2.22 puane

Uncertain etymology, although **pua-puane** may be glossed as 'glitter, sparkle, twinkle'. The term, which is roughly equivalent to AM 'murapati', is fairly consistently applied to the Amboina Cuckoo-dove, *Macropygia amboinensis,* and is possibly extented to other Columbidae (pigeons and doves), including *Gymnophaps mada.* **Puane** is ritually restricted for Sonawe-ainakahata, and in parts of west Seram features in a myth in relation to the sacred mountain Nunusaku [Ribbe, 1892: 179].

4.2.23 hutiene, manu pake

About 18 cm in length and described by Komisi as being a white version of **puane,** with black wing tips and an almost identical call. Probably *Ducula bicolor,* the white nutmeg imperial pigeon, and a species confined to coastal forest. The term may be extended to refer to other closely related pigeons. The call is said not only to resemble human speech, but actually to be speech.

4.2.24 (man) kihoke

Focal species is the purple-naped or black-capped lory, *Lorius domicella,* in AM 'kasturi kepala hitam'. This brightly-coloured bird is frequently kept as a household pet, and sold to outsiders. The category appears to be extended to other lories and related *PSITTACIFORMES*, including the red-breasted pygmy parrot, *Micropsitta bruijnii.* Ritually restricted for the clan Sonawe-ainakahata.

4.2.25 wekae

Two types are recognised:

4.2.25.1 weka uoi, weka noe msinae

(U)oi = 'call', presumably a reference to the scream of this parrot; **msinae** = red. Focal species is the red-sided eclectus parrot, *Electus roratus,* and identifications are generally of this species. It is equivalent to *Electus pectoralis,* the determination provided by Stresemann [1914: 89]. However, the term is occasionally extended to related parrots.

4.2.25.2 weka marae

Marae = 'blue-green'. Possibly the blue-eared lory, *Eos semilarvata.*

4.2.26 nakatua

Cognate with AM 'kakatua', literally **'kakak tua: elder sibling'.
Wekae** is sometimes spoken of as a type of **nakatua** (**nakatua wekae**),
which in some respects it may be said to resemble. Both are large with no
green plumage, which places them in Rand and Gilliard's [1967: 191-2]
Group 1 Psittacidae. Thus, the question frame 'how many kinds of **nakatua**
are there?', often meets with the response, 'two - **nakatua wekae** and **nakatua putie'. Nakatua putie** is the Moluccan cockatoo, *Cacatua moluccensis,*
distinguishable by its white body and salmon recurved crest (**orane**).
Cacatua galerita is known from east Seram and Seram Laut and differs from
C. moluccensis in its yellow crest.

The tail-feathers of **nakatua putie** are used by the Nuaulu for male
head-dresses, which like cockatoo crests are also known as **orane** [see Ellen
1993: plate 1.6e; and 11 here]. White cockatoos are caught live for pets or
for sale. Young birds are often caught by climbing trees and simply using
the hands, but noose traps may also be set. Ritually restricted for the clan
Sonawe-ainakahata.

4.2.27 konane

Consistently identified as the Moluccan or island parrot, *Tanygnathus
megalorhynchos affinis.* Informants described it as living in coastal low-
lands and feeding on the fruits of *Callophyllum inophyllum* (**auhutaune**),
manioc leaves, bananas and cane sugar.

4.2.28 kasituri

This Sepa word, though in AM 'kasturi', is not a Nuaulu term as such,
though it is heard sufficiently frequently to suggest that it is fast becoming
one. Nuaulu informants volunteered terms for four kinds of **kasituri: kasituri marae** ('blue'), **kasituri msinae** ('red'), **kasituri putie** ('white') and
kasituri masikune ('yellow'). These all refer to parrots and lories of the
order *PSITTACIFORMES*, and overlap in various ways existing Nuaulu
categories: **kihoke, wekae, konane, kunini, siseite** and **tuie**. On various
occasions informants identified *Charmosyna placentis* and *Eos (E. bornea
rothschildi, E. squamata, E. semilarvata) as* **kasituri,** although only one is
clearly and consistently linked with a specific terminal category, namely *Eos
squamata* with **kasituri msinae.** The most plausible explanation for these
terms and their referents are that the term **kasituri** was originally introduced
as a convenient collective term for parrots and lories prominent in trade,
with their obvious physical similarities. Binomials would then have been
the simplest way of differentiating between species highly variable in

PLATE 11: Moluccan cockatoo (**nakatua putie**:*Cacatua moluccensis*) kept as pet in Rohua: 7 August 1975 (neg 75-4-36).

coloration. The binomials are therefore best considered as rare residual terms for species for which no other name exists, and the uninomial as a convenient generic for grouping various small parrots.

4.2.29 kunin(i), (manu) punini (R. B.)

The similarity of **kunin** to AM 'kuning' ('yellow') may at first seem significant. Burkill, 1935 glosses Malay 'burung kuning' as *Oriolus*. Since there are no yellow orioles it would seem that in Malay the term also refers to the call of the bird, giving rise to an intentional or unintentional pun. Van Bemmel reports only one species from Seram, *Oriolus (bouruensis) forsteni*, and anyway available evidence at my disposal points rather to a focal *Eos bornea*, an equation consistent with Stresemann [1914: 81]. The Nuaulu distinguish two types: **kunini msinae** ('red') and **kunini marae** ('grue'). The latter is said to be slightly smaller. **Kunini** are sometimes kept by the Nuaulu as house birds.

4.2.30 siseite

Seite is a knife, perhaps an allusion to the sharp bill of this bird. A term applied to a large bright red (possibly meaning brown) fruit eating parrot, 'burung raja' in AM (R.B.). Stresemann [1914: 92] uses the term **siseite** to refer to *Alisterus amboinensis*, which certainly matches descriptions given by informants.

4.2.31 tuie

A possible onomatope, and root of the personal name Tuisa. Stresemann [1914: 90] uses the term **tuye** to refer to *Geoffroyus personatus*. According to more recent authorities *G. personatus* is not present on Seram and the term may therefore refer to the red-cheeked parrot *G. geoffroyi stresemanni*.

4.2.32 kukue

An onomatope - like English 'cuckoo' - which clearly identifies the species referred to as members of the order *CUCULIFORMES*. The focal species is almost certainly the oriental cuckoo, *Cuculus saturatus horsfieldi*, but the term is also used to refer to the grey-breasted brush cuckoo, *C. variolosus stresemanni*, the Malay bronze cuckoo, *Chalcites (Malayanus?) crassirotris* and *C. malayanus minutillus;* the Indian koel, *Eudynamys scolopacea orientalis*, and the channel-bill cuckoo, *Scythrops novaehollandiae*.

4.2.33 (manu) suanane

Possibly an onomatope alluding to the conspicuous song of this bird, though it is worth noting that **suanane** is 'marriage'. The focal species is the coucal, *Centropus bengalensis medius*, but the term may well be extended to refer to other Cuculidae. Coucals differ in important behavioural respects from cuckoos; seldom flying, and feeding on large insects and small vertebrates in shrubbery and grass. It is this which must lie behind the stark categorical contrast drawn by the Nuaulu between species which are otherwise morphologically close.

4.2.34 tuku-tuku

The reduplication and phonological characteristics of the term suggest an onomatope, although informants clearly identified it with specimens of owls (*STRIGIFORMES*), probably *Otus magicus*.

4.2.35 sakoa

Clearly the brown owl, *Ninox squamipila*.

4.2.36 kuna-kuna

A reduplicated term which suggests an onomatope. Possibly the long-tailed nightjar, *Caprimulgus macrurus mesophanis*. Stresemann [1914: 153] reports the term (rendered **guna-guna**) as being applied to *Corvus enca*.

4.2.37 man totane

The term is applied to the Papuan nightjar, *Eurostopodus papuensis*, although the species is not reported by Van Bemmel for Seram. The term may also refer to *Caprimulgus macrurus*, leaving **kuna-kuna** (4.2.36) as an onomatopoeic synonym. In 1973 the term was recorded as referring to a black and white plumed bird - a flying 'forest chicken'.

4.2.38 sioi, neune anae, man kina nahane

The first and most common of these terms is possibly onomatopoeic, in that it resembles closely the high-pitched whistle of the glossy swiftlet, *Collocalia esculenta esculenta*, to which the term is consistently applied. *Collocalia* is found in large numbers in caves frequented by Nuaulu in search of bats. All specimens identified were collected in mist nets on the Wakakau while Menzies and I were accompanying Nuaulu on a bat hunting party in July 1975. The term is almost certainly extended to include the congeneric *Collocalia spodiopygia*. Stresemann [1914: 150] lists **sioi** as referring to *Basilornis corythaix* (and also *Culomis*) but these attributions must surely be erroneous.

Collocalia is also sometimes referred to by the terms **neune anae** and **kina nahane**. **Neune anae** = 'child of the **neune**', this latter being the tree *Casuarina equisetifolia*, which also occurs as a personal name. **Kina nahane** was described to me in 1973 as having a white head and body, brown wings, with brown along the top of the wing- like a pigeon. One of the meanings of **kina** is stick, but it also occurs as a prefix in some insect names (e.g. **kinapari, kinapopote** and **kinapukune**). **Nahane** is literally 'rat/mouse', and also the hardwood *Mimusops elengi*, used, amongst other things, for the manufacture of barkcloth beaters.

4.2.39 tihalapeti

Consistently identified with the sacred kingfisher, *Halcyon sancta,* an Australian migrant which does not breed in the Moluccas. Said to be a bird of the ground rather than of the treetops. It is eaten if caught, but is not sought after. Stresemann [1914: 92] lists the term **tifanapeti** (which must surely be the same) for *Eurystomus orientalis.*

4.2.40 moi numte, manu numte, mui numte

On the meaning of **moi** see 4.2.11. **Numte** = 'mountain, far away'; also plant, see 2.2.4. This is certainly a kingfisher, possibly *Halcyon chloris chloris* or *H. lazuli,* or both. Stresemann [1914: 111] records the term **manu nute** for the crested swift, *Hemiprocne mystacea.*

4.2.41 (manu) tuamane

Tuamane is 'earth, ground, land, terrain'. The term is most usually applied to the dwarf kingfisher, *Ceyx lepidus lepidus,* although it may possibly be extended to the river kingfisher, *Alcedo atthis hispidoides.*

4.2.42 saneane

The common paradise kingfisher, *Tanysiptera galatea nais,* although the term may possibly be applied to other kingfishers. Stresemann [1914: 97] uses **saneane** to refer to *Halcyon sancta.*

4.2.43 sopite

The Papuan hornbill, *Rhyticeros plicatus plicatus,* in AM 'tuan-tuan'. This is a highly distinctive bird, on account of its horny bill crest (**pipane**), wide wingspan and the burring noise made when flying. It is sometimes shot for food, while its bill is occasionally used as a manufacturing material. The term **sopi sarane** would appear to refer to a sub-type although its range of application is quite unclear.

4.2.44 soihihi

A term which appears to be applied to flycatchers generally. On Seram this covers 13 species of the genera *Pachycephala, Monarcha, Myiagra, Muscicapa, Dendrobiastes, Eumyias, Muscicapula, Siphia* and *Rhipidura.* My only definite identification is for *Pachycephala pectoralis,* the golden whistler; also recorded as such by Stresemann.

4.2.45 tika poho-poho

Poho means to 'break into pieces', **tika** being 'very'; perhaps referring to the fact that these birds eat carrion. **Poho-poho** suggests an onomatope. Probably the Ambonese drongo, *Dicrurus bracteatus.*

4.2.46 opor

Opor is an onomatope for the call of this bird, which was rendered by informants as **'opor, opor...'**. Both **tika poho-poho** and **opor** suggest the only crow reported for Seram, *Corvus enca violaceus,* and both terms may also be extended to include *Dicrurus.*

4.2.47 (manu) tinnanae

Anae is 'child', **tin** a contraction of some other word, perhaps **tina** ('thunder'), giving the plausible translation of 'child of thunder'. This was a term applied to specimens of the black sunbird, *Nectarinia aspasia,* but the term must also include *N. jugularis clementiae* (ex *Cinnyris clementiae*), to which Stresemann [1914: 145] applied the term **tinenanae.**

4.2.48 kinoke

kinoke = 'dawn' (R.B.), in AM 'burung siang': *Philemon subcorniculatus.* The category probably also includes the endemic Seram honey-eater, *Lichmera monticola.* Sachse, 1907: 56 reports *Philemon* as being used by some Seramese in augury, although I have no evidence that this is the case for the Nuaulu. Stresemann, 1914: 26, 142 lists the term **ginoke** (which is presumably the same) for *Philemon.*

4.2.49 hiru

A possible onomatope. Described as a small green or yellow bird living in the forest bottom, no more than four meters from the ground. Eats fruit, grubs and larger insects.

TABLE 7 Distribution of Nuaulu bird terms according to order, family and number of known species on Seram.

Order	Family	Number of endemic species on Seram	Number of Nuaulu categories
CASUARIFORMES	Casuariidae	1	1
PODICIPEDIFORMES	Podicidedidae	1	0
PROCELLARIFORMES	Procellaridae	3	
PELECANIFORMES	Sulidae	2	
	Pelecandidae	1	
	Anhingidae	1	
	Phalacrocoracidae	2	1
	Fregatidae	2	
CICONIIFORMES	Ardeidae	11	1
	Threskiornithidae	2	1
ANSERIFORMES	Anatidae	4	1
FALCONIFORMES	Accipitridae	8	1
	Pandionidae	1	1
	Falconidae	5	1
GALLIFORMES	Megapodiidae	2	2
	Phasianidae	1	1
GRUIFORMES	Rallidae	6	1
	Jacanidae	1	
CHARADRIIFORMES	Charadriidae	4	1
	Scolopacidae	19	0
	Phalaropodidae	1	0
	Burhinidae	1	0
	Laridae	9	4
COLUMBIFORMES	Columbidae	15	2
PSITTACIFORMES	Psittacidae	11	8
CUCULIFORMES	Cuculidae	8	2
STRIGIFORMES	Strigidae	2	2
CAPRIMULGIFORMES	Caprimulgidae	1	2
APODIFORMES	Apodidae	3	1
	Hemiprocnidae	1	0
CORACIIFORMES	Alcedinidae	7	4-5
	Meropidae	1	0
	Coraciidae	1	0
	Bucerotidae	1	1
PASSERIFORMES	Muscicapidae	11	0
	Pachycephalidae	2	1
	Sylviidae	6	0
	Turdidae	3	0
	Maluridae	1	0

Nectariniidae	2	1
Corvidae	1	1-2
Dicruridae	1	1
Oriolidae	1	
Sturnidae	3	
Artamidae	1	
Laniidae	1	?3
Pittidae	2	
Hirundinidae	2	
Campephagidae	4	
Motacillidae	2	
Ploceidae	2	
Zosteropidae	5	
Dicaeidae	1	
Meliphagidae	5	1

4.2.50 katenane

Katehate is a strip of decorated red cloth constituting an important part of male ritual head-gear; **nane** is a sail. Term applied to a small passerine of uncertain identity.

4.2.51 suite

Term applied to a small passerine which flies low over the village in the evening with a shrill cry, described as resembling that of a bat (chapter 3.2.4).

4.2.52 sopate

Unidentified small passerine.

4.2.53 manu kasa neane

Kasa neane is literally 'thirsty, dried out'; an evocative term for this small black and blue bird. Said to be similar to **hiru**.

4.2.54 kura-kura

Possible onomatope. The reference is unclear.

4.3 Social and economic uses of birds

Some data relevant to this heading are located in the entries for individual categories. To summarise, the Nuaulu maintain that all birds may be eaten unless ritually restricted, and many are. Few species are specifically

FIGURE 6 Simple model of some aspects of the internal partitioning of the category **manue**

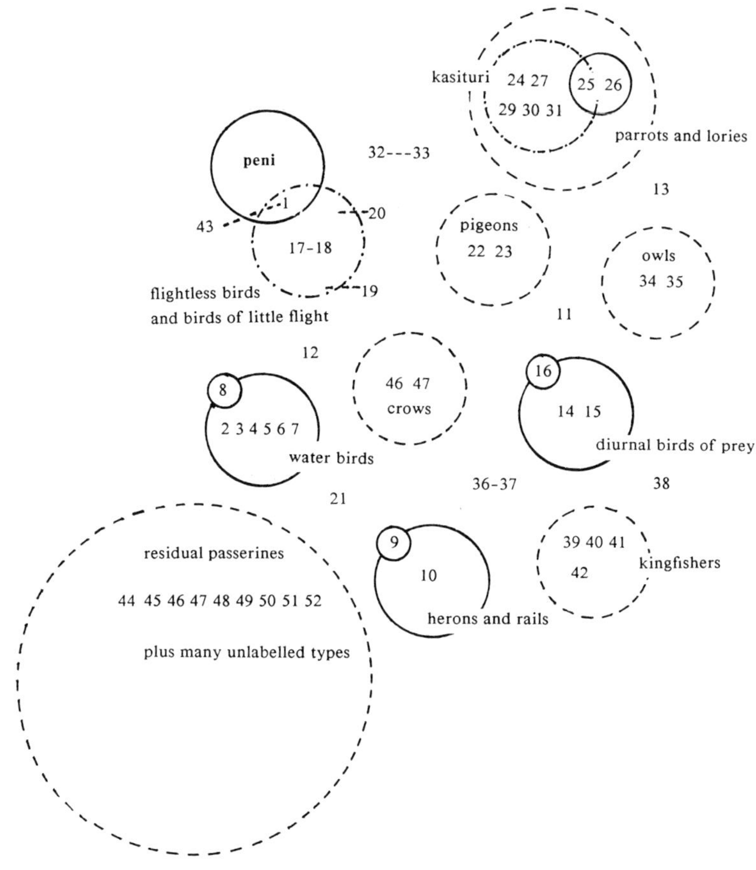

Legend:

______ indicates existence of unambiguous labelled intermediate or primary category

_ _ _ _ indicates existence of unambiguous covert category

_._._._. indicates existence of ambiguous semi-covert category

A continuous bar between numbers (e.g. 36-37) indicates unambiguous classificatory linkage; strong presumption of existence of covert category

A broken bar between numbers (e.g. 32---33) indicates a possible classificatory linkage.

Figures refer to reference numbers used in text. Unless clearly linked or enclosed, no particular significance should be attached to the spatial location of numbered categories.

hunted, and then for their plumage or for trade rather than for their meat. The eggs of cassowary, megapodes and chicken are occasionally eaten. *Casuarius* and *Rhyticeros* provide hard materials (bone, quills, bill and claw) used for making artifacts. A number of bird species are kept as pets by the Nuaulu, or (more frequently) are eventually sold in Sepa, to passers-by or in one of the other non-Nuaulu villages along the coast between Jalahatan and Masohi. This is particularly common for *Cacatua moluccensis* and *Lorius domicella*. Amir and Wind, 1978: 8 report that in the late nineteen-seventies live cockatoos could fetch between 3000 and 6000 Indonesian rupiahs for their hunters, and lorikeets between 2000 and 5000.

4.4 More inclusive categories for birds

In 50 percent of cases Nuaulu terminal categories for birds apply quite definitely to a particular species, and to no other (e.g. **nusi takanasi** = *Pandion haliaetus*). About 30 categories, however, are defined in relation to a particular focal species extended to other members of the same family. Where categories are identified in relation to focal species, such species obtain this position because of their salience, defined variously in terms of frequency of occurrence, size, economic significance, colourfulness, behavioural attributes and so on. Categories are then extended to include other related forms. In six percent of cases the extension may include members of different phylogenetic families, or European folk categories (e.g. **kukue**: *Cuculus saturatus* > other Cuculidae). Lumping of this kind appears to be understandably frequent with respect to some (though not all) migrants, occasional migrants and stragglers. In a further six percent of cases, categories appear to be defined in terms of a general prototype without a particular focal species (e.g. **nusi tanane**: all *PELECANIFORMES*, **muinu puane**: all Falconidae). Flexible use of terms is infrequent except (perhaps) in relation to kingfishers and certainly in relation to lories, where queries often prompted the introduced term **kasituri**, as a convenient inclusive, residual and otherwise covert category (figure 6).

Such wide use of the focal species and category extension strategy for groupings with a wide phylogenetic content implies a striking level of under-differentiation of avifauna by the Nuaulu, compared, say, with the Kalam [Majnep and Bulmer, 1977] and Tzeltal [Hunn, 1977]. This picture is evident from table 4.2, where under-differentiation of passerines in particular can be seen to be quite astonishing. Passerines are, of course, numerous, small and generally of little social significance. Indeed, the Nuaulu evidence justifies the view of passerines as a residual category of 'dikky birds', to use Hunn's term, in which there are only occasionally salient species for which we find consistently-applied and specific names.

The most general category for birds is **manue,** a term briefly glossed by informants as 'things which fly'. In its most inclusive sense this term is rendered **manua panesi,** or **manua pusita,** meaning 'all of those things which fly'. The inclusion of bats (the term **manu notane** is heard) in such a grouping is variable, but invertebrates are always excluded. In practice the term **manue** is applied both to '**manua** which fly' and **manua** which do not. **Manua** which fly are sometimes called **manua roe ai atu** (lit. 'those **manua** above the tree tops'), or **manua roe naue.** Flightless **manua** (cassowaries and scrub hens) are called variously **manua tuamana, 'manua** of the ground', **manua poe tuamane (poe** = prep. 'down'), or **manu nohu (nohue** = n. 'under'). In *The Cultural Relations of Classification (1993).* I note that cassowaries are regarded as having 'lost their wings' (chapter 6.7), so one possibility is to define the category **manu** in etic terms as vertebrates that fly or are regarded as having once flown.

The commonest use of the term **manue** is in this final sense where it corresponds to a concept of 'birdness' closely resembling that of folk English; that is animals possessing bills (**hohai, supente**), wings (**kihene**), feathers (**hunue, man hunue**) and claws (**kanikura**). While it is impossible to present a single Nuaulu model for the internal arrangement of **manue,** some idea of widely-shared linkages and sub-categories are indicated in figure 6.

Notes to Chapter 4

1   There are specific verbs in Nuaulu language to refer to the call of the cockerel and hen which are vaguely onomatopoeic: **'tunkeku'** and **'erekota'** respectively, where **ere** is a pronominal prefix for non-human actors.

CHAPTER FIVE

TURTLES

Reptiles are treated here in three separate chapters, reflecting the three broad groupings recognised by the Nuaulu themselves: testudines, saurians and snakes. The still rudimentary state of our herpetological knowledge of Seram, and of the Moluccan islands in general, has been commented upon elsewhere [Ellen et al, 1976a]. However, the information available for reptiles is probably more complete than that for amphibians. Reptiles are also more numerous, and prior to 1969 43 certain species had been reported, plus one uncertain provenance for the Elaphid snake *Aspidomorphus muelleri*.

Of the total corpus of known species, 19 were definitely observed during fieldwork, and most of them obtained as specimens in the Nuaulu area. However, there is evidence that the Nuaulu are familiar with at least 39 reptile species. The difference between this figure and the number known zoologically from the entire island may be accounted for by relative geographical distribution, particularly differences between highlands and the coast.

5.1 The turtles of south central Seram

There are just two families of testudines on Seram: the Cheloniidae, of which there are now four species recorded, and the Emydidae, of which one species is found. The leatherback turtle, *Dermochelys coriacea,* previously unrecorded for Seramese waters, was observed in 1973. A checklist of testudines reported from south central Seram is presented in table 8.

5.2 Nuaulu categories applied to turtles

5.2.1 peku, sahaunue

This is the only testudine category elicited for which specimens were actually collected, all other species being either not observed at all or too large to preserve and transport under normal ethnographic field conditions. **Peku** corresponds to *Cuora amboinensis,* the freshwater turtle, and is consistently identified as such by informants. Its morphological distinctiveness compared with other known turtles and the absence of an extensive testudine fauna makes this understandable. It is much smaller than the marine turtles with a shell length of little over 20 cm. It has a domed carapace, hinged plastron and unlike the marine turtles can completely withdraw into its shell when threatened. It is a herbivorous inhabitant of ponds and marshland.

TABLE 8 Checklist of testudines for the Nuaulu area of south central Seram

Species	Ecological zones					Nuaulu glosses
	1	2	3	4	5	
CHELONIA						
Emydidae - freshwater turtles						
Cuora amboinensis Amboinan box terrapin	-	-	-	+	-	**peku**
Cheloniidae - sea turtles						
Dermochelys coriacea leatherback turtle	-	-	-	-	+	**enu ikae**
Eretmochelys imbricata hawksbill turtle	-	-	-	-	+	**enu hunane**
Chelonia mydas green turtle	-	-	-	-	+	**enu hunane**
Caretta caretta	-	-	-	-	+	**enu hunane**

Key. Zone 1 = above 1000 m, principally montane rain forest; zone 2 = tropical lowland rain forest; zone 3 = secondary rain forest, garden and village areas; zone 4 = freshwater and swamp forest; zone 5 = marine and estuarine.

Peku is a primary totem for the clan Nepane-tomoien, who are prohibited from using this name for the animal, although this attitude by no means applies to all totems. Instead it is referred to as **sahaunue,** the term also used for the coconut shell or AM 'tampurung'. Literally, it has an affinity with **msaha,** a term used to refer to or address married men before they have become fathers, or **sahane**, a generic term for affines; **unue** = 'head'. The connection between the literal and referential meaning of the term **sahaunue** is obscure. It is not the same as that used for carapace, which is **nonia.** **Peku** is sometimes referred to by individuals from clans for which it is not a totem as **peku sahaunue.** Such synonyms indicate the importance of correlating names with actual specimens, rather than informants' descriptions or ethnographers' observations, since different names have a tendency to suggest separate categories.

5.2.2 Enu (penu)

Enu refers to marine turtles of which two types are recognised: **enu ikae (ikae** = 'fish') and **enu hunane (hunane** = 'moon'), also known as **enu tipope,** the meaning of which is unknown. The first refers precisely to the leatherback turtle *Dermochelys coriacea,* although Stresemann, 1927: 61 reports that in the closely related language of Hatue the same species is referred to as **tipope.** There are no testudines in museum collections from Seram apart from *Cuora amboinensis,* although *Chelonia mydas, Eretmochelys imbricata* and *Caretta caretta* are known for Ambon, and these species must certainly be present in Seramese waters. Sachse, 1907: 57 reports *Chelonia (i.e. Eretmochelys) imbricata* from Seram. Descriptions of **enu hunane** appear to be referring to *E. imbricata,* but the label is probably applied by extension to all other marine turtles that are obviously not **enu ikae.** *Chelonia mydas* has a maximum shell length of just under 1.5 m and varies from light to dark brown, sometimes with a tinge of olive. The scutes are marked with radiating or mottled darker markings. *Caretta caretta* has about the same maximum length and is usually reddish-brown in immature specimens. It can be distinguished from *Chelonia mydas* and *E. imbricata* by the five (instead of four) costal scutes on each side of the carapace. It also has a relatively larger, broader, head. *E. imbricata* is the smallest of the **enu** species, with a maximum shell length of one meter. The overlapping scutes on its shell (in all but the very largest) distinguish it from all other marine turtles. The carapace is amber with streaks of various shades of brown and yellow.

5.3 General remarks on Nuaulu turtle classification

On the basis of observation, tests and interviews, testudines are clearly treated by the Nuaulu as a separate natural group, although there is no generic term for them, such as AM 'tuturuga'. The possession of a distinctive morphology is backed-up by the assertion that **peku** may develop from **enu** eggs. This belief in inter-species ontogeny is a widespread characteristic of Nuaulu ethnozoology and is discussed in detail in chapter 6.4 of *The Cultural Relations of Classification.* If an **enu** (type unspecified) lays 150 eggs in the forest - I was told by Saniau - 50 of these would become **peku** on reaching the shore. There is no evidence that either **enu** (which is more likely) or **peku** is used as a generic for testudines in the way **notu** is used for frogs (see 8.4). The distinction made within this covert group is probably simply habitational, between marine (**enu**) and freshwater (**peku**), although size (large:small) was also offered by informants as a distinguishing characteristic. This provides us with the taxonomic structure in figure 7.

FIGURE 7 Nuaulu classification of turtles arranged as a taxonomic hierarchy

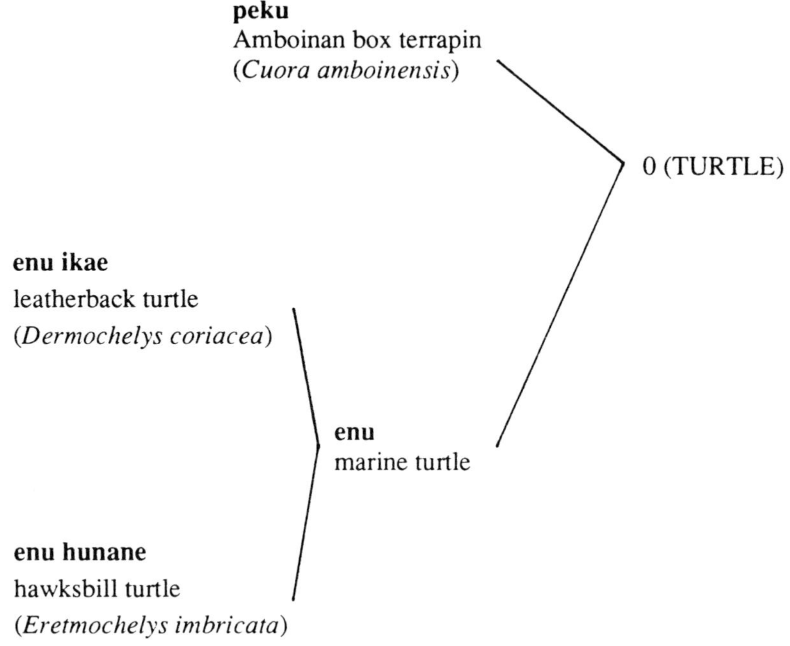

by extension:

   green turtle (*Chelonia mydas*)
   loggerhead turtle (*Caretta caretta*)

Specimens of *Cuora amboinensis* were consistently identified as **peku**. **Enu** of any kind are rarely encountered and there is probably some difficulty among younger Nuaulu in differentiating between **enu hunane** and **enu ikae**. This is partly to be accounted for by the fact that Nuaulu do not generally engage in marine fishing, and it is even forbidden for some clans e.g. Sonawe-ainakahata. No doubt this is connected with the fact that prior to resettlement the Nuaulu were located in the upper valleys of the Nua and Ruatan rivers, with little opportunity to see - let alone eat - sea turtles. There is no evidence though that the term **enu** is a recent loan word, which might seem a reasonable inference in the circumstances.

As far as can be judged, there is little or no variation between informants in terms of assigning terminal categories for turtles to more inclusive groups. Although this domain is not named, there is no difficulty in representing its inner relations as a taxonomy. This is partly due to the limited content and distinctive morphology of the group. In the card-sorting test, 22 informants out of a sample of 25 grouped **enu ikae** and **peku** together. These were grouped with other reptile categories as follows (numerals indicate number of informants):

	puha	puo$_1$	isa	imasasae	poso$_4$	kasa'un
	crocodiles	monitors	sail-tails	geckos	skinks	*Calotes*
peku	11	2	2	1	1	1
enu ikae	9	4	4	3	2	2

The preference for grouping with **puha**, and then **puo**$_1$ and **isa**, appears to be due to the fact that like turtles, these are to a certain extent, aquatic beasts. There seems to be no consistent reason for associating turtles with the remaining categories. The **enu/peku** distinction itself is probably made largely on the basis of habitat (sea:land).

5.4 Social uses of turtles

Except for Nepane-tomoien, turtles and their eggs may be eaten by anyone, and it is not even entirely clear whether the taboo with respect to **enu** is rigidly enforced for this clan. However, they are not preferred foods and occurred nowhere among the items consumed during a dietary survey undertaken in 1970-71. They are not sought after or hunted, but readily eaten when available. Although *E. imbricata* is occasionally eaten, the flesh can be poisonous and its consumption has been known to cause human deaths. The shell of hard-backed marine turtles (**en(u) unte** = ' **enu** skin'), usually that of *Eretmochelys imbricata,* is fashioned into finger rings (**sopao**) and small anklets (**niti anae**), generally for children. [1]

Peku is undoubtedly the most commonly consumed turtle, and occurs in large numbers around the mouths of the larger creeks flowing into the Banda sea, such as the Upu and Pia. The larger marine turtles are only eaten when caught on the beaches, usually when laying eggs. A large (80 kg) **enu ikae** was caught in this way in August 1973. When the shell, calipee, dung, blood and waste had been discarded just over half of the original weight was available for eating. However, the odour of the leatherback is disliked, and compared to the smell of decomposition - **haue kopue** (**haue** = 'smell', **kopue** = 'rotten, decaying, foul').

Note to Chapter 5

1   For example, see Museum of Mankind (London) Reg. Nos. 1972 AS1, 224-7; Rijksmuseum voor Volkenkunde, Ellen field catalogue No. 551; British Museum (Natural History) Reg. No. 1976.933.

CHAPTER SIX

LIZARDS AND RELATED FORMS

6.1 The lizard fauna of south central Seram

The order Crocodilia is represented by a single member of the family Crocodylidae. Five families of lizards (Sauria) occur: four species of geckos (Gekkonidae), three distinct species of the family Agamidae, one species of monitor lizard (Varanidae), and eleven species of skinks (Scincidae). A checklist of lizards and related forms reported for south central Seram is presented in table 9. Species identification compared with Nuaulu categories applied to actual specimens collected are set out in table 10.

6.2 Nuaulu categories applied to crocodiles and lizards

6.2.1 puha

The term is probably cognate with AM 'buaja'. Crocodiles (*Crocodylus porosus*) are more or less confined to the estuarine zones of the larger rivers of south central Seram, such as the Ruatan, Jala and Lahati, though occasionally they may be found in coastal waters. They are a large carnivorous inhabitant of both fresh and saltwater and may reach a length of nearly 6 m. The young feed on frogs and invertebrates while adults eat larger animals including man. Crocodiles have been reported by informants from other localities and appear to have been more widely distributed in the past. They are not actively hunted, although they may have been formerly and will still be killed if the opportunity arises. The crocodile is a totemic animal for the clans Matoke, Sopanani, Sonawe-aipura and Huni, though for none of these does it appear to be a primary totem. In the case of Matoke it is believed to have 'evolved' from one of the founders of the clan. It is clearly feared and gives its name to a type of taboo sign or scare charm (**wate puha**), which threatens the victim with a fate worse than (but including) death, by being eaten alive. There is some evidence to suggest that the fearsome reputation of the crocodile and the fact that its increasing inaccessibility prevents it being hunted make this an animal respected and tabooed by all Nuaulu. This is not made clear in any formal sense, but was the considered view of certain informants.

6.2.2 puo

A term applied to all monitor lizards (*Varanus indicus*), commonly found along the coast, particularly in coconut groves, and known in AM as 'sua-sua' (plate 12). At 1.5 m it is the largest lizard on the island. It differs from the others by its long slender forked tongue. It is a good swimmer and is often found in the vicinity of water. It has a distinctive coloration of numerous small yellow spots on a blackish or dark olive ground. In Nuaulu nomenclature this sexually dimorphic species is divided into two terminal categories: **puo inae (inae** = 'mother')(♀) and **puo pipane** (♂). There are no difficulties in identifying *Varanus* as **puo,** although youngsters some- times find difficulty in sexing them.

Puo is a primary totem for the clan Matoke, on whose war shields it is sometimes depicted. It is believed to be descended from the dog of the ori- ginal ritual head of this clan [Ellen 1993: table 6.2] As the clan Matoke is guardian of the village ritual house (**suane**), a **puo** motif is sometimes carved on the large drum attached to this structure. **Puo** is a secondary totem for the clans Penisa, Pia and Nepane-tomoien. Like **puha, puo** also gives its name to a type of scare charm.

Monitor meat is commonly eaten by clans other than Matoke, and after the reticulated python it is probably the most important reptile species appearing in Nuaulu diet. It is not normally actively hunted, but will be caught if located in coconut palms near the village. As a pest which will eat chickens and ducks, it is often more important to destroy it in the interests of pest control than to capture it for food. Although the Nuaulu appear to have no use themselves for the skin at the present time, this is always carefully removed as it can be later sold to dealers trading between Seram and Ambon, where it is used as an abrasive, an ingredient in Chinese medicines, and possibly also for other purposes.

6.2.3 isa

This is a term applied to the agamid lizard *Hydrosaurus amboinensis,* the sail-tailed lizard, so-called because the crest, which is very distinct on the back, becomes even higher on the base of the tail, especially in mature males. It may exceed a meter in length, of which two-thirds is tail. It is unusual among lizards in being almost entirely vegetarian. It is semi- aquatic, living on the banks of rivers and streams, and is regarded as a pest. It is occasionally eaten, but not especially sought after.

Like *Varanus, Hydrosaurus amboinensis* is a sexual dimorph and the category **isa** is divided by the Nuaulu into two terminal categories: **isa inae** = 'mother')(♀) and **isa pipane** (♂). **Pipane** refers to the serrated dorsal

TABLE 9 Checklist of lizards and related forms recorded for the Nuaulu area of south central Seram.

Species	Ecological zones					Nuaulu glosses
	1	2	3	4	5	
CROCODILIA						
Crocodylidae						
Crocodylus porosus	-	-	-	+	+	**puha**
estuarine crocodile						
SAURIA						
Gekkonidae - geckos						
Hemidactylus frenatus	-	-	+	-	-	**imasasae numa**
common house gekko						
Gekko vittatus	-	-	+	-	-	**imasasae ai ukune**
Agamidae - dragon lizards						
Calotes cristatellus	-	-	+	-	-	**kasa'un**
Draco lineatus amboinensis	-	-	+	-	-	**hohone**
Hydrosaurus amboinensis	-	-	+	+	-	**isa**
Varanidae - monitors						
Varanus indicus	-	-	+	+	-	**puo**
water monitor						
Scincidae - skinks						
Tiliqua gigas	-	-	+	-	-	**nopa inae**
Mabuya multifasciata	-	-	+	-	-	**poso noha kunie**
many banded skink						
Mabuya 'rudis'	-	-	+	-	-	**poso ai totu kopue**
						poso noha metene
Carlia fusca	-	-	+	-	-	**poso ai totu kopue**
Carlia sp. (prob. *fusca*)	-	-	+	-	-	**poso ai totu kopue**
Dasia smaragdina moluccarum	-	-	+	-	-	**poso kaimarane**

Emoia cyanura	-	-	+	+	-	**poso kaimarane**
Emoia kuekenthali						
notomoluccensis	-	-	+	-	-	**poso kaimarane**
Eugongylus rufescens	-	-	+	-	-	**nopa hanaie**

Key. Zone 1 = above 1000 m, principally montane rain forest; zone 2 = tropical lowland rain forest; zone 3 = secondary rain forest, garden and village areas; zone 4 = freshwater and swamp forest; zone 5 = marine and estuarine.

sail. The male is the larger of the two, and one specimen examined measured 75 cm from head to tail. **Isa** does not appear to be of any totemic or other sacred significance. Because it is less common than **puo** it is correspondingly more difficult to identify consistently and certainly to sex with any degree of accuracy. Because of its superficial similarity to *Varanus,* it is often described as 'a kind of **puo**'.

6.2.4 hohone

A term consistently applied to the flying lizard, *Draco lineatus amboinensis,* although it was also applied by three informants on one occasion to specimens of *Calotes cristatellus.* This may represent a generic usage or may be simply informant error in responses from young children. However, the abundant distribution of this unmistakable species in garden areas near the village would seem to make confusion unlikely.

Nuaulu recognise several different terminal varieties of flying lizard though - I think - with no particular consistency or categorical significance. Napwai isolated **hoho ai ukune**. **Ai** = 'tree, woody shrub'; **ukune** = 'end, top': **ai ukune** = 'treetop, tree branches, far forest', as in AM 'ujung kayu'; also **aikune** (= 'trunk, base, beginning': R.B.). In addition **hohone** are sometimes described as either **hoho metene (metene** = 'black, dark') or **hoho marae (marae** = 'blue-green'). Of these, only **hoho metene** was actually used with reference to collected specimens, and it seems that flying lizards which are not of a particularly dark hue are simply described as **hohone,** and lexically differentiated no further. **Hohone** is not normally differentiated into these various types, and it appears that the terms distinguish variant individuals, rather than terminal categories in the strict sense of the word.

TABLE 10 Species identifications compared with Nuaulu categories applied to 108 lizard specimens

	imasasae ai ukuna	imasasae numa	kasa'un	hohone	puo	poso	poso noha kunie	poso ai totu kopue	poso kaimarane	poso kaimarane onate	nopa hanae	Number of informant responses	Number of specimens
Gekko vittatus	1	-	-	-	-	-	-	-	-	-	-	1	1
Hemidactylus frenatus	1	50	-	-	-	-	-	-	-	-	-	51	42
Calotes cristatellus	-	-	21	3	-	-	-	-	-	-	-	24	14
Draco lineatus	-	-	-	5	-	-	-	-	-	-	-	5	5
Varanus indicus	-	-	-	-	7	-	-	-	-	-	-	7	3
Mabuya multifasciata	-	-	-	-	-	7	17	3	-	-	-	27	21
Mabuya 'rudis'	-	-	-	-	-	2	2	-	-	-	-	4	4
Carlia fusca	-	-	-	-	-	2	1	3	-	-	-	6	5
Carlia sp *	-	-	-	-	-	-	-	2	-	-	-	2	2
Dasia smaragdina moluccarum	-	-	-	-	-	-	-	-	2	1	-	3	2
Emoia cyanura	-	-	-	-	-	-	-	-	9	1	-	10	7
Emoia kuekenthali notomoluccensis	-	-	-	-	-	-	-	-	-	-	-	0	1
Eugongylus rufescens	-	-	-	-	-	-	-	-	-	-	1	1	1
Totals	2	50	21	8	7	11	20	8	11	2	1	141	108

* Full identification not yet available

6.2.5 kasa'un (kasa uni)

All specimens of *Calotes cristatellus* were consistently allocated to this category (but see 16.2.4). This species is extremely common in garden areas and around villages. It is a great favourite among children when playing, the lizard being tied around the belly with a rattan lead. Although **kasa'un** is clearly grouped with **puo, isa** and **hohone** in terms of morphological features, Komisi says that it is distinguished from other animals by the

PLATE 12:Naunesi Sonawe-ainakahata with monitor lizard (**puo**) shortly after it had been caught in a coconut palm outside Rohua, 13 July 1975 (neg. 75-1-21).

length of its tail (which may be as much as three times the length of the body). Because of this it is said to be derived from **tekene** ('snakes'), although this is not evident from their classificatory position.

6.2.6 (n)imasasae

The term is applied to all geckos (Gekkonidae) known to the Nuaulu, and as might be expected, it is the pentadactyly and the suction pads on the digits which are the crucial distinctive features of the category. Geckos are also distinguished by the small number of eggs which they lay, compared with other lizards.

There are two terminal categories applied to geckos: **imasasae (ai) ukune** and **imasasae numa.**

6.2.6.1 imasasae (ai) ukune, imasasae ai atu

As has already been mentioned (6.4.2), **ai ukune** (but also **ai atu**) refers to 'tree top, tree branches, far forest'. In this context it distinguishes between species of geckos on the grounds of habitat, between forest and village. It is described by informants as being an habitue of gardens, banana and coconut palms.

Although only two specimens were subsequently collected and formally identified, it is clear that this term is more-or-less consistently applied to *Gekko vittatus,* despite the fact that one of the specimens so identified was *Hemidactylus frenatus.* On this occasion there appears to have been a genuine error on the part of the collector, partly due to the fact that the specimen did not seem to fit into **imasasae numa,** on account of its light coloration and mottling. This is a good example of a small sample of specimens and responses badly reflecting general classificatory practice.

Imasasae (ai) ukune is normally physically distinct from **imasasae numa** on account of its greater size (with an adult length of 25 cm) and colour. It is brown above and below with a distinct cream stripe along the middle of its back. This stripe bifurcates on the neck, the two branches running along the sides of the head to the eyes. The original tail has about four cream bands, but regenerated tails lack obvious pattern. The dorsal granules are interspersed with numerous small tubercles. It is the largest gecko on Seram.

Saniau says that there is only one 'natural kind', but did distinguish varieties on the basis of the predominant hue of the skin: **imasasae (ai) ukune metene, imasasae (ai) ukuna putie** and **imasasae (ai) ukune msinae** (= 'black', 'white', and 'red' respectively). One informant described it as being derived from **puo.**

6.2.6.2 imasasae numa

This gecko lays its eggs in the thatch of Nuaulu houses (**numa** = 'house'). Of the 50 responses elicited from informants for 41 specimens, all applied the term to *Hemidactylus frenatus,* the common South-East Asian house gecko. This has an adult length of 12.5 cm, is greyish or pinkish brown above with or without indistinct darker mottling. A dark-edged light streak passes on the sides of the head and through the eye. The skin on the back is soft and smooth without tubercles. The Nuaulu see no necessity to recognise sub-types. Since this is the most common gecko found in south central Seram, the term **imasasae** refers to this species when used in an unqualified way.

As for the Kalam [Bulmer et al, 1975: 299-300], so among the Nuaulu, familiarity with the house gecko tends to breed contempt. They are worthless little animals of slight ritual and no economic significance and yet are found everywhere in the village. Gecko-chasing is a common pastime among children who catch their tails and treat them as unfair game in play-hunting with bows and arrows, despite their small size. Yet geckos are curious creatures, and may often out-wit humans in their efforts to catch them. Their tails are easily detachable and may come in a variety of shapes and sizes; some are even forked. Komisi said that **imasasae numa** (and by extension probably all geckos) are derived from the eggs of **puo** (6.2.2), and card-sorting tests revealed a consistent association between these two categories although there is no morpho-syntactic indication of this relationship. Although not utilised totemically, **imasasae numa** has a particular significance for the clan Somori. If one is heard calling in a house where there is a sick person this is an indication that the spirit of a recently deceased member of the clan (**sionata**) has entered the body of the gecko and is guarding the patient. If the gecko is then caught it must in no way be harmed.

6.2.7 poso

This term is used in a maximum of four senses:

1. all Sauria
2. all Sauria that are either geckos or monitors
3. skinks (Scincidae)
4. skinks other than **nopa hanaie** and **nopa inae**

The sense indicated is generally clear from the context. Its broadest sense (1) is rarely encountered except in the artificial situation of an ethnographer asking abstract questions concerning the classification of animals. Its second sense is more common and includes all skinks plus the Nuaulu categories **kasa'un** and **hohone.** Whether **nopa hanaie** and **nopa inae** should be included here or whether they are *sensu stricto* to be included in a more restricted sense (3) is variable. The narrowest sense (4) can almost certainly be accorded the status of a primary category, contrasting with **imasasae, kasa'un, hohone, puha, isa** and **puo.** Sub-divisions of the primary category are usually indicated morpho-syntactically by further differentiation (e.g. **poso noha kunie,** where **noha kunie** has no meaning in terms of skink classification except when preceded by **poso**). The different senses of the term **poso** are examined further below.

6.2.7.1 poso noha metene

No informants identified specimens as **poso noha metene (metene** = 'black, dark'), although it is possible (on the basis of informants' descriptions) that four unidentified specimens of *Mabuya 'rudis'* are to be placed in this category. On Seram this skink appears to be acting as a species distinct from *Mabuya multifasciata,* rather than simply as a sub-specific variant.

6.2.7.2 poso noha kunie

Of the 18 responses indicating this category (**kunie** = 'tumeric', *Fibraurea chloroleuca*), 17 referred to specimens of the many banded skink, *M. multifasciata.* Larger individuals are said by informants to be coprophagous. In general, they are regarded as dirty and polluting.

6.2.7.3 poso ai totu kopue

This name may be glossed 'rotting tree leaf skink' (**ai** = 'tree, wood'; **totue** = 'leaf'; **kopue** = 'rotting, decaying'). It is described by informants as living on decaying leaf litter and on middens. Its colouring and patterning provide an appropriate camouflage. Of eight specimens assigned to this category, each with a single identification, three were *M. multifasciata,* three *Carlia fusca* and two *Carlia sp.* As there are no other species of the genus *Carlia* known from Seram, it is possible that specimens of *Carlia sp.* will turn out to be *Carlia fusca.* This would confirm the most plausible view that the type characteristics of **poso ai totu kopue** are those of this latter species.

6.2.7.4 poso kaimarane

The term for this skink is sometimes simply rendered as **kaimarane** (**kaie** = 'bright', **marane** = cuscus, (chapter 2.2.1). It is said to be able to climb trees, unlike other members of the primary category **poso.**

Out of 13 responses for nine specimens, two indicated *Dasia smaragdina moluccarum,* and nine *Emoia cyanura.* An unidentified specimen of *E. kuekenthali notomoluccensis* is also probably assigned by the Nuaulu to this category. Nuaulu appear to sub-divide it still further (which would give it an intermediate status), but not consistently so. Two identifications indicated **poso kaimarane onate (onate** = 'large'), which appears to refer to *Dasia smaragdina moluccarum,* commonly found on and around coconut palms. Other members of this category seem to be simply referred to as **poso kaimarane.** One informant volunteered the term **poso kaimarane ikine** (**ikine** = 'small'), but this appears to be hardly ever used. Anyway, it is clear that it refers to *Emoia cyanura.* This is a distinctive little creature seen commonly on bushes, fallen trees and in the gardens, although it is found in particularly large numbers (almost to the exclusion of other lizards) in the great sago swamp forest on the Ruatan river. The species is sexually dimorphic, the female (though smaller) being the most easily recognisable with its bright blue luminous tail and yellow body with black horizontal banks. This sexual dimorphism is not recognised terminologically by the Nuaulu, although Retau'une did say that there were three morphological types of **poso kaimarane: onate,** presumably *Dasia smaragdina,* and what were clearly the two sex types of *E. cyanura.*

6.2.7.5 nopa(i) hana(i)e, nopa(i) ina(e)

These terms, rendered **nopa hanaie/inae** and **poso nopa hanaie/inae** with equal frequency, distinguish two terminal types (**hanaie** = 'male', **inae** = 'mother') of a covertly recognised intermediate category. On morpho-syntactic grounds, one might expect them to be members of a category labelled ***nopane,** but there is no evidence that this is current practice. **Nopa asu** means 'to hold a dog' and **nopa okum** 'to hold one's nose'. It may be that 'to grip, to hold' is indicated here, referring to the reputation this animal has for biting humans. **Nopa hanaie** is described as very large, with 'a stomach as big as a human fist'. Its head is said to be more like that of a snake than of a skink, and by this is presumably meant that the jaws are well-developed and stand out laterally. Its dorsal surface is described as having the coloration of **poso,** while its belly is white with yellow flanks. It is said to have a vicious bite which is capable of killing a man, and the recent case of Hunimora Nepane-tomoien was related to by Komisi as an instance of this. Its distinctive cry is known as the **kako nione,** and is said to be audible in the late afternoon. The identification of a single specimen indicates **nopa hanaie** as *Riopa (Eugongylus) rufescens.*

Nopa inae is unlikely to be the female of *R. rufescens* as the species shows no obvious sexual dimorphism. It is probably *Tiliqua gigas*, which is much larger, has darker limbs, a relatively heavier head, and some similarities in coloration. The skin of **nopa inae** is said to be similar to that of the death adder *Acanthophis antarcticus* (chapter 7.2.9), with pink and black vertical stripes, and with which it is sometimes confused. Though not regarded as poisonous and never having been known to lead to death, the bite can cause an infection.

These carrion-eating skinks are the only ones regarded by the Nuaulu as being edible.

6.3 Variation in the identification and classification of lizards and related forms

In any estimate of the consistency of Nuaulu reptile identifications crocodiles must be ignored. They are seldom encountered and were not collected as specimens. Vivid folk descriptions and their occurrence in stories, however, suggests that most adult individuals would have little difficulty in recognising them for what they are, except perhaps in the case of immature specimens.

Lizards present a much more complex interpretative picture than either turtles or crocodiles. Specimens of *Gekko, Hemidactylus, Varanus, Draco, Calotes, Emoia, Dasia* and *Eugongylus* appear to be more or less consistently identified from what evidence is available (table 10). The reasons for this appear to be, at least in the first instance, largely a matter of their individual morphological distinctiveness. The sub-specific discriminations of *Varanus indicus* (**puo**) and *Hydrosaurus amboinensis* (**isa**) are due to sexual dimorphism and appear to be consistently made. The sub-specific discriminations of *Draco lineatus* (**hohone**) and *Gekko vittatus* (**imasasae (ai) ukune**) are probably made rather inconsistently and relate to distinctions in location, behaviour, and in certain cases, subtle colour differences. They are probably not regarded by the Nuaulu as being 'natural kinds' in the sense of being sexually self-reproducing.

In contrast, however, there is a rather special problem in determining the identity of certain skinks. Specimens of *Carlia fusca, Mabuya multifasciata* and *Mabuya 'rudis'* were all identified by informants in a relatively inconsistent way (table 10). If it is assumed that *Carlia sp.* is *C. fusca*, then the terms **poso noha metene, poso noha kunie** and **poso ai totu kopue** can be seen as being applied according to two relative criteria- degree of darkness and habitat- such that they overlap the three species *Mabuya 'rudis', M. multifasciata* and *Carlia fusca* as shown in figure 8. It should be noted that the

FIGURE 8 Relationship between Nuaulu categories for certain skinks and their phylogenetic content.

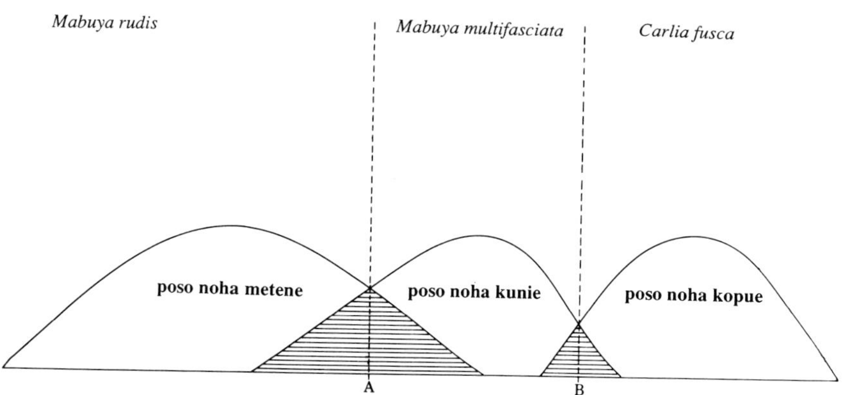

degree of overlap between **poso noha metene** and **poso noha kunie** (A) is greater than that between **poso noha kunie** and **poso ai totu kopue** (B). This reflects the phylogenetic distance between the corresponding biological species. Although they seem to indicate the three phylogenetic species involved in terms of their ideal type characteristics, the three indigenous categories appear to be sometimes applied in a relative sense, as in A is to B (labels) as *x* is to *y* (observed animal). What may be labelled **poso noha kunie** in contrast to another labelled **poso ai totu kopue,** may in contrast to a lighter coloured individual of *M. multifasciata* be spoken of as a **poso noha metene.** This kind of labelling, where relationships between observed animals is more important than equivalences between actual specimens and categories defined in terms of absolute distinctive features, occurs elsewhere in Nuaulu ethnozoology, but is nowhere better exemplified than with reference to skinks, and possibly frogs. In both cases the use of 'loose labels'[c.f. Hunn, 1976: 518] appears to be related to the social non-utility of the species. The remaining **poso** categories do not display this property and this is undoubtedly related to their morphological distinctiveness.

6.4 More inclusive categories for lizards

There is no single term for crocodiles and lizards, although they are certainly construed as being closely related forms.

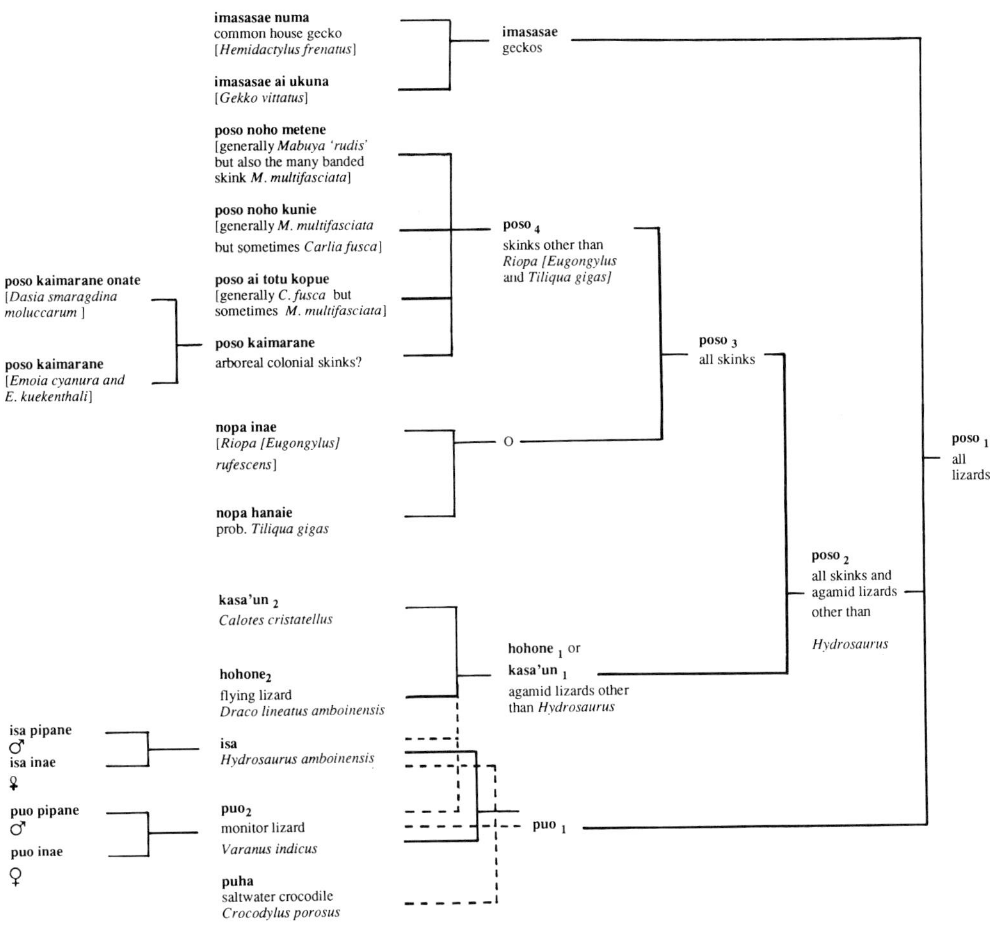

FIGURE 9 Nuaulu classification of lizards arranged as a taxonomy

Puha is sometimes spoken of as a type of **puo** and **puo** is sometimes spoken of as a type of **poso.** Although one might argue on syllogistic grounds that **puha** is therefore a type of **poso,** in my experience it is never classified as such by the Nuaulu and some informants emphatically denied the possibility. In the card-sorting test **puha** was grouped most frequently (in order of descending frequency) with **isa** (15), **puo** (12), **peku** (11), **poso** and **enu ikae** (9 each). It is clear that there is some ambiguity here between **puha** classified morphologically (as a saurian related to **poso, isa** and **puo**) and behaviourally (as an amphibious or aquatic form related to **enu ikae** and **peku**). Although, on balance, the 'morphological' classification appears to be the more important, even this is not completely clear. One informant justifying this schema commented 'all are **puo,** they live in water or sometimes enter water'. It is indeed the case that *Hydrosaurus* and some monitors are semi-amphibious.

The card sorting test revealed very similar patterns for puo_2 (1) and **isa**:

	puo_2	isa	imasasae	$poso_4$	puha	$kasa'un_2$
puo_2	-	23	20	18	12	5
isa	23	-	17	15	15	5

The strong association between **isa** and **puo** confirms what has already been said about their relationship. More unexpected in view of the above discussion is the relatively low degree of association with **puha,** though in this case crocodiles may be seen as aquatic in contrast to the more terrestrial monitors and sail-tail lizards. Whether an animal is aquatic, arboreal, terrestrial, or whatever, is clearly very much a question of relativity.

The patterns of association for **kasa'un** and **poso** are as follows:

	imasasae	isa	puo_2	$poso_4$	puha	$kasa'un_2$
$kasa'un_2$	8	5	5	5	1	-
$poso_4$	21	15	18	-	9	5

The low grouping of both categories with **puha** again suggests classificatory distancing on the grounds of terrestrial-aquatic distinctions. The strong grouping of $poso_4$ and **imasasae** is almost certainly because both are characteristically inhabitants of the village, as opposed to the forest. $Poso_4$ and $kasa'un_2$ are distanced classificatorily, and this must certainly be related to their morphological distinctiveness, the village-forest dichotomy and perhaps also the contrast between 'polluting' and 'clean'.

What is noticeable immediately from these groupings is that they cannot be arranged in terms of a neat taxonomy. A possible classification of Nuaulu crocodile and lizard categories arranged as a taxonomy is set out in

FIGURE 10 Types of variation in Nuaulu classification of lizards.

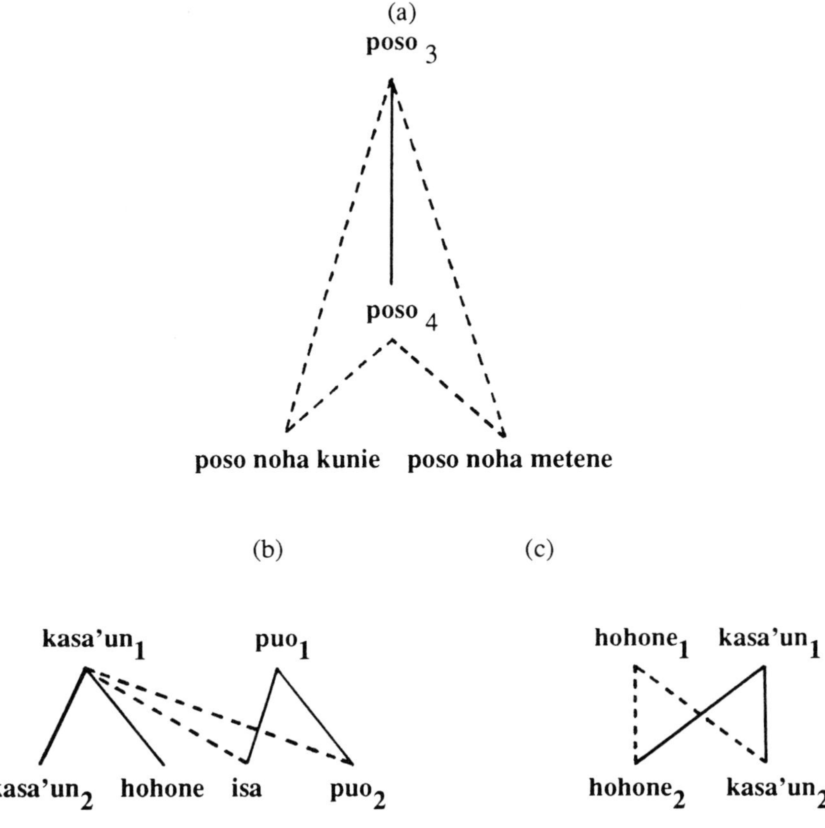

figure 9 based on the maximum number of classificatory 'levels' elicitable. This is inevitably no more that a convenient representational device and does not include all the possible connections suggested by the results of the card sorting test. No one informant volunteered such a schema, and in this sense it is that of an 'ideal speaker-hearer'. It is also important to see it as the *maximum* number of distinctions; it is unlikely to be elicited in this form from any hypothetical average informant. Some levels are commonly omitted and the degree to which they are relevant depends very much on context. For example, the terminal categories included in $poso_4$ may also be seen as types of **poso₃** without the intervention of **poso₄** and the covert category (0). That is, the condition of transitivity is understood (figure 10a). Then there are cross-cutting classifications. For example, in response to the question 'what kinds of **kasa'un₁** are there?', two informants listed **hohone₂, kasa'un₂, isa** and **puo₂** although the last two are also commonly grouped together as **puo₁** in contrast to **kasa'un₁** (figure 10b). It is interesting to note that in the card-sorting test one fifth of all informants grouped **isa** and **puo₂** with **kasa'un₂**. Finally, there is the variable substitution of terms in a class inclusive relationship (not indicated in figure 9). Thus, in one elicitory context it may be in order to speak of **kasa'un** as a type of **hohone,** and in another of **hohone** as a type of **kasa'un** (figure 10c). This usage is also reported for **isa/puo, nopa hanaie/nopa inae** and (for amphibians) **poro-poro/notu.** Nuaulu use of the conjunction **nai** to mean 'of the order of' (as in **isa nai puo**) appears to be only used when the terms are not normally related morpho-syntactically. We would therefore expect this to be a common usage in type C variation.

CHAPTER SEVEN

SNAKES

7.1 The snake fauna of Seram

There are four families of snakes known from Seram: Typhlopidae (four species), Boidae (four species), Colubridae (five species) and Elaphidae (four species). Of these, six species were definitely recorded during fieldwork, and obtained as specimens in the Nuaulu area. There is evidence that the Nuaulu are familiar with between ten and 16 species.

A checklist of snakes reported for south central Seram is presented in table 11. Species identification compared with Nuaulu categories applied to actual specimens collected are set out in table 12.

7.2 Nuaulu categories applied to snakes

There are at least 20 Nuaulu terminal categories applied to snakes strictly defined, and all can be included within the primary category **tekene.** In referring to forms within this category, the term is almost always contracted to **teke** (e.g. **teke tuamana**). Snakes as a whole are totemic for the clans Somori (primary totem) and Nepane-tomoien (secondary totem), although in practice respect varies from species to species, being most important for **teke patona.**

7.2.1 teke tuamana

Members of this category are said by informants to be small burrowing snakes living under the ground (**tuamana** = 'ground', 'earth'). From informant's descriptions they are clearly worm snakes (Typhlopidae). No specimens have been obtained to date and they appear to be rarely found in the immediate vicinity of Nuaulu settlements. Four species are reported from Seram by earlier workers, three of the genus *Typhlops* and *Ramphotyphlops multilineatus.* The Nuaulu recognise no sub-division of the category **teke tuamana,** and since it is of no special significance ethnozoologically, and its precise contents difficult to track down, this is not entirely surprising. Indeed, the biological differences between the species are limited to some fine points of anatomy.

7.2.2 teke patona

The term **patona** (= AM 'patola') is used to describe a type of imported woven cloth important in certain rituals. **Patona** specifically refers to the pattern on such cloths, which is said to resemble the markings of the snake, or vice versa [c.f. Jensen, 1939: 372][1]. Barbosa reports its presence in Banda as early as 1518 and describes it as a cloth of silk or red linen with a special pattern [Rouffaer and Juynboll, 1914: 417]. Among other groups, such as the Christian Ambonese, it is part of bridewealth payments and fines payable in cases of adultery or murder [Cooley, 1962: 28, 30, 37].

The category corresponds precisely to *Python reticulatus,* the reticulated python, and is identified consistently as such by informants. Generally regarded as the longest of snakes, this species is commonly found in villages and adjacent areas. The larger pythons, of five or six meters in length, are normally restricted to nearby lowland forest areas, and it is these creatures that are of economic significance for the Nuaulu. Reticulated pythons are unquestionably the most economically important reptiles, and were the only ones to appear in a sample dietary record made for a four month period in 1970[2]. The gall bladder (**monie** = 'bile', 'gall bladder') is often sold to non-Nuaulu as medicine. Burkill reports that it is used by Malays as a draught for callous ulcers [Burkill, 1935: 1848]. It is also probably part of the overseas Chinese pharmacopoeia, where it is used as a cure for bloody diarrhoea, other disabling haemorrhages and malarial miasmas [Schafer, 1967: 216-7] The Nuaulu have no traditional use for it. The meat is some-times cooked with fat (which is considered a delicacy) in lengths of green bamboo stoppered with leaves of *Languas speciosa* (**wainite**). More generally it is roasted using a rapidly constructed rack of saplings and green bamboo, about 60 to 100 cm off the ground.

Teke patona is the primary totem of the clan Somori, although avoidance behaviour is generally extended to other snakes, with the possible exception of **mainase** (7.2.8). It is one of the most notably autochthonous animals known to the Nuaulu, and, according to more than one informant, **teke patona** (with its many eggs) has given rise to **nopa inae** (one large egg), **poso** (two eggs) and the centipede, **niniane** (two eggs). On another occasion, a second informant added **kasa'un, hohone, isa, puo** and **imasasae** to this list. It is possible that the length of this list represents bias due to the clan affiliation of the informant being Somori.

TABLE 11 Checklist of snakes recorded in the Nuaulu region of south central Seram.

Species	Ecological zones					Nuaulu glosses
	1	2	3	4	5	
Typhlopidae - worm snakes						
Typhlops braminus	-	?	-	-	-	teke tuamana
Typhlops kraali	-	?	-	-	-	teke tuamana
Typhlops ligorostris	-	?	-	-	-	teke tuamana
Ramphotyphlops multilineatus	-	?	-	-	-	teke tuamana
Boidae - pythons and boas						
Morelia amethistina	?	+	+	-	-	teke soata
amethystine python						
Python reticulatus	?	+	+	-	-	teke patona
reticulated python						
Candoia carinata	-	+	+	-	-	mainase
Pacific boa						
Colubridae - colubrid snakes (harmless)						
Dendrelaphis pictus pictus	-	+	+	-	-	teke tam niane, teke konomete
colubrid snakes (mildly venomous)						
Boiga irregularis	-	+	+	-	-	teke panarine
banded tree snake						
Chrysopelea rhodopleuron	-	?	-	-	-	teke msinae
freshwater snakes						
Fordonia leucobalia	-	-	-	+	-	teke waene
Cerberus rhynchops	-	-	-	+	-	teke waene
Elapidae - elapid snakes						
Acanthophis antarcticus antarcticus	-	+	+	-	-	nanate
death adder						
Aspidomorphus muelleri	-	+	+	-	-	?nanate
sea snakes						
Laticauda colubrina	-	-	-	-	+	teke nuae
Laticauda laticauda	-	-	-	-	+	teke nuae

Key. Zone 1 = above 1000 meters, principally montane rain forest; zone 2 = tropical low-land rain forest; zone 3 = secondary rain forest, garden and village areas; zone 4 = fresh-water and swamp forest; zone 5 = marine and estuarine.

TABLE 12 Species identifications compared with Nuaulu categories applied to snake specimens.

	teke patona	mainase	mainase metene	teke tam niane	teke konomete	teke panarine	nana putie	Number of informant responses	Number of specimens
Python reticulatus	5	-	-	-	-	-	-	5	5
Candoia carinata	-	7	1	-	-	-	-	8	7
Dendrelaphis pictus	-	-	-	8	2	-	-	10	2
Boiga irregularis	-	-	-	-	-	8	-	8	3
Acanthophis antarcticus	-	-	-	-	-	-	3	3	1
Total	5	7	1	8	2	8	3	34	18

7.2.3 teke (m)sinae

Msinae = 'red'. There is very little information available on this category, but it appears to correspond to *Chrysopelea rhodopleuron*.

7.2.4 teke panarine

This snake is described by informants as being small but long (the size of a small **teke patona**) and tree-dwelling. A common habitat is said to be the leaf bracts of the sago palm. It is not, however, to be confused with what in AM is called 'ular sagu' (sago snake), the grub of the sago weevil (*Rhynchophorus bilineatus*). The confusion is compounded by the fact that this snake is sometimes called 'ular saja' in AM.

Opinions on coloration vary from 'black and red', 'grey back, yellow body', 'red and yellow' to 'blue'. Despite such conflicting descriptions, it appears that the banded tree snake (*Boiga irregularis*) is being referred to; a fanged species of notoriously variable colour, but generally thought not to be dangerous to humans. The term **teke panarine** was given in all eight responses to requests to identify three specimens of *B. irregularis*.

7.2.5 teke ai atu

Ai atu, 'treetop'. The term is used in a generic sense to mean 'tree snake'. Used in this way, one informant distinguished two kinds: **teke panarine** and **teke soata,** referring respectively to *B. irregularis* and to *Morelia amethistina*. It is possible that **teke ai atu** also refers to the mangrove snake, *B. dendrophila*. This snake is widespread throughout Indonesia, at least as far east as Sulawesi. It is strikingly banded in black and yellow and grows considerably larger than *B. irregularis*. This might more easily account for the remark about size, but without recorded specimens or more definite evidence it remains speculation.

7.2.6 teke tam niane, teke konomete

Tam niane can be glossed as 'return to the village'. The reason for this name is that whenever this snake is seen lying across a path the traveller must return to the village. With the clan Somori this rule applies to most snakes. It is quite clear from examining specimens, observation and interviews that **teke tam niane** refers to the immature form of *Dendrelaphis pictus*. **Teke konomete** refers to the mature form of *Dendrelaphis pictus*. The general prohibition on travel signalled by the appearance of **teke tam niane** is not extended to sightings of snakes in this form.

7.2.7 teke soata

Described as a large 'black and yellow' tree snake of the forest, as opposed to the village periphery and garden areas. According to one person, 'it can grow as thick as a human thigh'. It is sometimes hunted and eaten. The meat is separated from the fatty parts and roasted in much the same way as reported for **teke patona.** The fat is cooked separately as a relish and eaten with various kinds of root tubers, or used as cooking fat. One specimen identified as *Morelia amethistina*.

7.2.8 (teke) mainase

This snake is distributed widely in the Nuaulu area and is commonly found anywhere from house rafters to mature forest. It clearly corresponds to the genus *Candoia (= Enygrus)*. The Nuaulu recognise two types of **mainase: mainase putie** (= 'white') and **mainase metene** (= 'black'). Five specimens (seven responses) of *Candoia carinata* were identified consistently as **mainase putie.** One specimen was identified as **mainase metene.** These terms appear to be simply identifying colour morphs within the species.

Mainase is an interesting category in that its classification as **tekene** is somewhat ambiguous. The prefix **teke** is not indissolubly part of the term morphosyntactically, and taboos concerning snakes in general (as for the clan Somori) apply less strictly. They are sometimes ignored altogether. In justifying this practice, one individual suggested that it was not a 'true' snake. The reasons for this are unclear, but morphological differences may be relevant. *Candoia* is relatively short and fat compared with most other snakes known to the Nuaulu, the shape of the head appears to be somewhat different, and it is generally regarded as being relatively harmless. If a person is bitten by a **mainase,** the appropriate therapy is to immediately find and eat some chicken faeces.

7.2.9 (teke) nanate, mnanate

Nanate is the Nuaulu name for the Muslim village of Sepa, within the jurisdiction of which all Nuaulu villages lie. Known in AM as 'ular bisa' (= 'poison snake'), referring to the most important characteristic attached to it by the Nuaulu, this snake is regarded as the most dangerous of its kind, being able to bite and climb the body. The fear of **nanate** is exploited in its use for a type of scare charm (**wate**) by the clan Penisa. Bites are usually treated by tying a cloth, string or other tourniquet above the bite and placing a silver or alloy ring (**sopa nanate**) on the wound. No invocation is involved. The ring may be plain, but characteristically features a stylised human skull. The source of these charm rings is unknown but they were probably (and may still be) manufactured by smiths in Amboina, or even further afield. The constant wearing of the ring is said to afford protection. Rings of this type are also associated with **teke panarine,** but documentation concerning their use is less extensive.

In view of the fearsome reputation of **nanate,** it is curious that the clan Nepane-tomoien has a ritual relationship with it similar to that which the clan Somori has with the gecko. That is, if one is heard calling in the house of a sick person it indicates that an ancestral spirit has entered the snake's body to protect that person. The same rules of respect apply here as apply for the gecko.

There are three types of **nanate** (using the term in its narrow sense) recognised by the Nuaulu. First there is **nana metene** (= 'black'). This is also known as **nana mnakasopa unu,** referring to the way in which this snake 'lifts its head off the ground to look round' (**unu** = 'head'). This behavioural trait appears to be characteristic of long thin snakes, but uncharacteristic of short fat ones - such as **mainase.** This type of **nanate** is also described as having a smaller head than other types. Secondly, there is **nana putie,** the 'white' **nanate,** and thirdly, **nana msinae,** the 'red' **nanate. Nana putie** corresponds precisely to the death adder, *Acanthophis antarcticus.* I am inclined to think that both **nana putie** and **nana msinae** are simply variants of this species. With its smaller head, **nana mnakasopa unu** may be *Aspidomorphus muelleri,* the existence of which on Seram has yet to be confirmed.

The term **nanate** is also sometimes used in a broader sense to include the lizards **nopa hanaie** and **nopa ina** (6.2.7.5), a practice which is discussed further below.

7.2.10 teke waene

Appears to refer to freshwater (**waene**) snakes, although also sometimes extended to anguillid eels. Two species of freshwater snake are recorded from Seram, *Fordonia leucobalia* and *Cerberus rhynchops rhynchops.* Specimens of neither were collected.

7.2.11 teke nuae

Nuae = 'sea'. These snakes are sometimes found in rock pools at low tide. The ventral surface is black and white, the dorsal surface differing according to variety. Many kinds of seasnake occur in seramese waters, although *Laticauda* is the only genus which habitually comes on shore. Of this genus, *L. colubrina* and *L. laticaudata* are reported.

7.3 Classifying the contents of the category 'tekene'

In its broadest sense the category includes all snakes (SERPENTES). In addition it is sometimes extended to include certain affiliated categories for lizards, invertebrates and eels (figure 11).

The invertebrate categories are **niniane** (centipedes), **nikenuke** (millipedes), **tuaman nae** (earthworms) and **ai ntone** (certain perichaet worms), all of which are described in more detail at chapter 14.1.5-9. In a card sorting test, **nikenuke** and **tuaman nae** were consistently classified with **tekene** (100 percent of a sample of 25 informants). This in itself does not necessarily mean that they are always regarded as **tekene,** but rather that the perceived resemblances between snakes, earthworms and millipedes are closer than between these animals taken together and other reptiles and amphibians. However, they are regarded by some informants quite definitely as **tekene,** although the failure to use the prefix **teke** on any occasion only confirms their rather anomalous position. Additionally, apart from the covert 'worm' category already mentioned, these invertebrates are not even classified as a separate group in contrast to other kinds of **tekene** [c.f. Bulmer, 1968: 629-33].

There are four other somewhat anomalous categories broadly grouped together as **tekene: mainase, nanate, teke nuae** and **teke waene**. **Mainase** has already been shown to be anomalous on morphological grounds and because of its relative harmlessness compared with most other snakes. **Nanate,** however, appears to be singled out for precisely the opposite reason - that it is the most harmful snake known to the Nuaulu. For this reason also it is often grouped with **nopa hanaie** and **nopa inae,** the most feared of lizards. Curiously, both **mainase** and **nanate** are sometimes grouped together as living on the ground rather than in the trees, under the ground or in water. **Teke nuae** is clearly anomalous because of its marine habitat, the morphological adaptations undergone by sea-snakes, and because of their similarity to certain eels. **Teke wane** may also be an ambiguous term for some Nuaulu, though of course many snakes will swim in freshwater if necessary and this would not seem to present a classificatory problem.

Eels present a special problem. The eel categories **awane** and **yapato** are systematically dealt with here in chapter 9, with other fish (**ikae**). Indeed, some informants were quite emphatic that eels were not **tekene**. Somori, for whom snakes are prohibited food, will eat eels; while Pia, for whom **awane** is prohibited, will eat snakes. On the other hand, a few informants concluded that perhaps eels were snakes which had become fish, and others still claimed that they definitely were snakes. The binomial **teke yapato** is sometimes heard, perhaps in part because because some seasnakes resemble moray eels quite closely.

Taken in its broadest sense, the category **tekene** can be conveniently sub-divided into a core group of 'true snakes' and a peripheral group of 'snake-like forms'. This is evidently in part morpho-syntactically motivated, but is supported by general classificatory evidence, as well as by explicit statements from informants. When the term **tekene** is used to include both groups, it is sometimes qualified as **tekene panesi** (= (in this sense) 'all', 'all of them', 'entirely' c.f. Indonesian 'segala') or **tekene pusirei** (= 'all of them', things).

'True snakes' are those consistently prefixed by the term **teke** (e.g. **teke patona, teke tuamana...**). In morphological terms they are distinguished from lizards by the shape of the head, the tongue, the length and shape of the body and (perhaps most importantly) leglessness. Behaviourally, 'true snakes' shed their skin, possess a distinct form of locomotion, are generally

FIGURE 11 Contents of the Nuaulu category **tekene** arranged as a taxonomy. The broken lines indicate variably attributed affiliations.

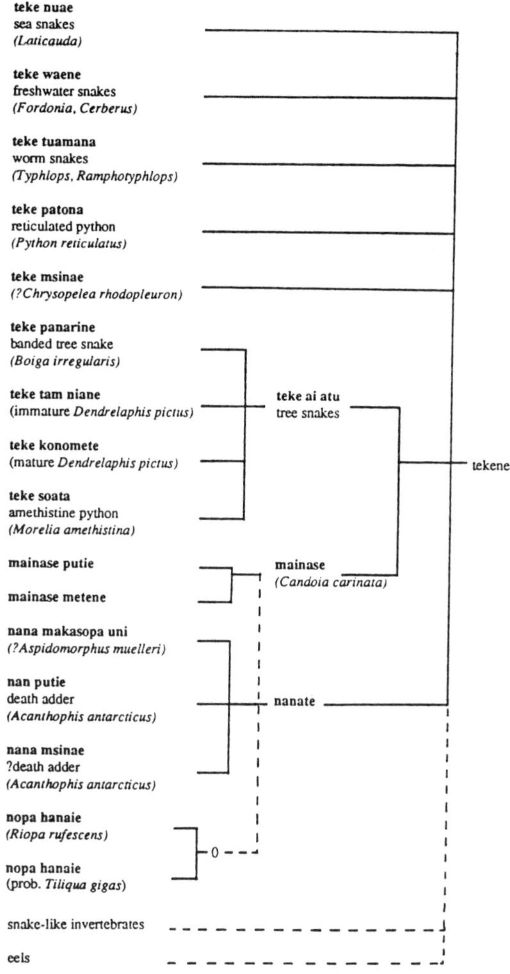

terrestrial and harmful to humans. By contrast, 'snake-like forms' are allocated to the primary category **tekene** on the basis of a varying assortment of morphological and behavioural criteria, but never possess the combination of characteristics which distinguish 'true snakes'. In some cases, they may be allocated to the category on the strength of one or two criteria only. **Tuaman nae** and **ai ntone** are problematic on most grounds except that they have long flexible bodies, snake-like locomotion and are legless. **Nikenuke** and **niniane** even possess legs. **Awane** and **yapato** are excluded from 'true snakes' partly on the grounds of being aquatic, but perhaps also because of their overall fishiness. It is only to be expected that the peripheral grouping has grey areas of overlap between other primary and more inclusive categories, in both indigenous and phylogenetic terms.

I shall comment here on just two of the criteria of 'true snakes', leglessness and harmfulness, simply to indicate the problems surrounding the use of such criteria. Although I am satisfied as to the folk significance attributed to leglessness in defining **tekene,** one informant insisted on one occasion that snakes do - in fact - have legs, used when climbing trees. This seems to be a reference to the two hemipenes of the snake. He justified this statement by pointing out that giant millipedes (**nikenuke**) have countless legs and yet they are regarded as **tekene.** This represents an interesting case of an informant justifying an 'erroneous' observation as to the characteristics of 'true snakes' by reference to 'snake-like forms', a group set aside precisely because of the absence of this particular diagnostic characteristic.

On the question of harmfulness, it is important to distinguish animals which can recognizably cause harm from those that are feared. The two groups are not necessarily coterminus: harm is concerned with the objective behaviour of such animals, fear is bound up with an often complex set of attitudes. It is not my intention to explore Nuaulu avoidance and phallicism of snakes in detail here. What data are available suggest that fear of snakes (and lizards) versus other groups is not as marked among the Nuaulu as it is for the Kalam [Bulmer et al, 1975:300-4].

The Nuaulu distinguish in general terms between snakes that are harmful - **teke kahatene** (= 'dangerous') and those that are harmless - **teke iake** (= 'good'). Sometimes a particularly dangerous snake is specifically termed **teke kahatene** (in AM, 'ular setan', demon snake). This is most commonly applied to the death adder, which explains the partial synonym **nana sakahatene** (see section 7.2.9). **Sakahatene** is a type of evil spirit, and use of the term indicates a belief that it is 'guarding the jaws' of the creature. The slaughter of such an animal is said to bring great problems. For one informant:

> If it is possible to trace the snake to its lair - which is difficult - the hole will be found to be closed up - because it is a **sakahatene.**

Harmfulness is therefore a characteristic of 'true snakes', although not all snakes are as harmful as others. **Mainase** is not a 'true snake', partly on morphological grounds, but partly because it is regarded as harmless. On the other hand, **nanate** is not a 'true snake' because it is regarded as being exceptionally dangerous. Clearly, while harmfulness is a characteristic of 'true snakes', extremely harmful ones are accorded a special classificatory status. In the case of **nanate** it means that it is lumped with other reptiles for which there is a marked aversion - **nopa hanaie** and **nopa inae** (figure 11).

7.4 Variation in the classification of the contents of the category 'tekene'

Nuaulu identifications of snakes appear to be on the whole consistent. The kinds of variation found in Nuaulu classification of the contents of the category **tekene** are of three basic types:

1. variation as to whether certain forms are to be included or excluded from the category **tekene** in its widest sense;

2. variation as to whether certain categories are to be included or excluded from the category of 'true snakes'; and

3. incidental variation stemming from the omission or incorporation of particular degrees of inclusiveness, as with the examples given for the classification of lizards and crocodiles.

Category allocations which are not universally shared by adult Nuaulu are depicted with a broken line in figure 11. The shape of the classificatory structures elicited must therefore be expected to vary, both between different informants and for the same informant on different occasions and in different contexts.

Variation of the first type applies to six terminal categories, covering **awane, yapato, nopa hanaie, nopa inae** and the invertebrates. None of these typically include phylogenetic snakes, though **awane** and **yapato** may occasionally do so. The status of the invertebrates has already been commented upon.

The only other point which must be made with respect to this first type of variation relates to the unusual position of **nopa hanaie** and **nopa inae**. These categories were never described to me as **tekene,** although they were on occasions said to be **nanate**. Since **nanate** are definitely said to be **tekene,** one might argue on syllogistic grounds that **nopa hanaie** and **nopa inae** are therefore, in fact, **tekene;** but there is absolutely no evidence to suggest that this is how the Nuaulu view the matter and some definite evidence that they would eschew such an approach. Nuaulu ethnozoology suggests that syllogistic logic is not a fundamental part of folk classification as it is in

Linnaean taxonomy.

The second kind of variation applies to the status of the following categories: **teke wane, teke nuae, mainase** and **nanate.** The anomalousness of these categories which accounts for such variation has already been discussed.

The only significant example of the third type of variation concerns the broad use of the term **mainase** to refer to both **mainase** and **nanate:**

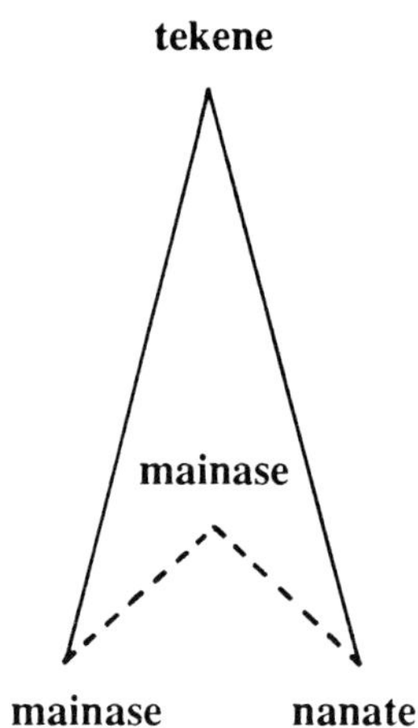

It is so because of the otherwise contrasting features of these two lower-order categories, and because (apart from **teke waene** and **teke nuae**) they are the only other SERPENTES in the peripheral category. The same kind of variation undoubtedly occurs with respect to **teke tam niane/teke konomete,** where the variable element is the degree to which the covert category '0' is recognised. Finally, the degrees of inclusion indicated by the terms **teke ai atu** and **tekene** (narrow sense) may sometimes also be omitted. The reason for the significance of the **mainase/nanate** variation is that here the intermediate term **mainase** (wide sense) brings together categories that are usually quite distinct. In the other cases the relationship between the terms is normally understood, although an intermediate classificatory grouping may be sometimes omitted.

Notes to Chapter 7

1   Gonda, 1973: 95-6 reports that in Malay and other Indonesian languages **patola** has a cluster of meanings on the theme of 'gaily coloured cloths'. It is evidently derived from Sanskrit where it refers to both a silk double ikat [Barnes, 1989: 19, following Bühler], originally Gujerati, and to an edible gourd, *Trichosanthes* spp. According to Burkill, 1935: 2178 this latter usage also occurs in Malay as **petola ular,** 'snake gourd'.

2   **Teke patona** was eighth in importance in a list of 12 sources of animal protein: 150 grams was consumed per adult head in two households during the survey period giving 0.67 per day (0.01 gram of protein, 8.82 Cals.) [Ellen, 1975b: 138].

CHAPTER EIGHT

AMPHIBIA

8.1 The amphibian fauna of south central Seram

The amphibian fauna of Seram is still imperfectly understood, despite the efforts of those working on various aspects of Moluccan natural history in the first half of the present century [Rubenkoning, 1959: ii]. Even Rumphius, who in other respects was a notable pioneer of Moluccan zoology, seems to have paid little attention to it [Greshoff, 1902; Wit, (ed.) 1959].

Frogs are the only amphibians known from the Moluccas. Of more than 20 Anuran families recognised, only three seem to occur on Seram. Prior to 1969, six species had been reported: *Rana grisea ceramensis* and *Platymantis papuensis* from the family Ranidae, *Phrynomantis fusca* from the family Microhylidae, *Litoria infrafrenata infrafrenata*, *L. amboinensis*, and *L. vagabunda* from the family Pelodryadidae. All of these were obtained between 1970 and 1975 in the south central part of the island, with the exception of *Rana grisea* and *L. vagabunda*. *R. grisea ceramensis* is known only from a single specimen collected in central Seram at about 1000 m in 1919, while *L. vagabunda* is also rare, being known from only two specimens, collected in 1872, one from Seram and one from New Guinea. In addition, *Rana modesta* was found (previously unknown from Seram), plus two hitherto biologically undescribed species: *Litoria sp. (bicolor* group) and *Rana* sp.

The species most commonly encountered by the Nuaulu is *Litoria infrafrenata,* a noisy green tree frog common in coastal and village areas and among the largest of its kind. This frog, however, has a wide altitudinal distribution, certainly occurring at 700 m, and possibly higher. It has a snout vent length of up to 13.5 cm, is immaculate green above with a green throat and whitish venter. There is a distinctive white stripe along the upper lip extending onto the sides of the neck. The hands and feet are webbed.

All other species, although present and audible in the immediate Nuaulu area, are less obtrusive. *Litoria amboinensis*, *Phrynomantis fusca*, *Platymantis papuensis* and *Litoria sp. (bicolor* group) are certainly present along the coastal strip. *L. amboinensis* is of medium size, with a snout vent of up to 6.5 cm. It is grey-brown above with irregular darker markings, is white below and has a throat spotted with brown. In contrast to the Pelodryadidae, *P. fusca* has fairly short hind legs, a little longer than the head and body together. The fingers and toes are unwebbed with tips that are only slightly expanded (again contrasting with the Pelodryadidae). It is rather large for a microhylid with a snout vent length of about 5 cm. It is dark grey-brown above, flecked more or less profusely with white, and is whitish below. *Rana modesta* is only reported from sago swamp forest towards the mouth of the Ruatan. The Ranidae on Seram are broad-headed, with a relatively pointed snout, large eyes and distinct tympanum. Their hind legs are long, much longer than the head and body together. Tips of the fingers and toes are slightly enlarged. The feet are webbed but the hands are unwebbed. *R. modesta* is a medium-sized frog with snout vent length of up to 7 cm. It is brownish olive above with or without a medium lighter stripe, whitish below and has a throat mottled with brown. *Rana* sp. was only obtained at the village of Piliana at an altitude of 700 m to the north of Teluti Bay.

A checklist of amphibians reported from south central Seram is presented in table 13.

TABLE 13 Checklist of amphibians recorded in the Nuaulu region of south central Seram.

Species	Ecological zones					Nuaulu glosses
	1	2	3	4	5	
Pelodryadidae						
Litoria infrafrenata	-	+	+	+	-	**poro-poro, notu**
Litoria amboinensis	-	+	+	+	-	**notu, inararai**
Litoria sp. (*bicolor* group)	-	+	+	+	-	**notu anae**
Ranidae						
Platymantis papuensis	-	+	+	+	-	**notu, kere, teteye**
Rana modesta	-	-	-	+	-	**notu, kere**
Rana sp.	-	+	+	+	-	**notu**
Microhylidae						
Phrynomantis fusca	-	+	+	+	-	**notu, kako**

Key. Zone 1 = above 1000 m, principally montane rain forest; zone 2 = lowland tropical rain forest; zone 3 = secondary forest, garden and village areas; zone 4 = freshwater and swamp forest; zone 5 = marine and estuarine.

Although Seram certainly has a depauperate fauna [Ellen, 1978b], it is unlikely to be quite so depauperate as current knowledge suggests. Most of the material collected is very widespread and quite typical of cultivation zones. It is probable that a more thorough investigation would reveal further endemic species.

8.2 Nuaulu categories applied to frogs

There are seven Nuaulu terminal categories applied to frogs. These are listed in table 14 against biological species to which they correspond.

TABLE 14 Species identifications compared with Nuaulu categories applied to 53 frogs.

Species	notu	notu anae	poro-poro	kere	teteye	inararai	kako	Total number of responses	Total number of specimens collected 1970-1975
Litoria infrafrenata	1	-	11	-	-	-	-	12	12
Litoria amboinensis	6	-	-	-	-	1	-	7	6
Litoria sp. (bicolor group)*	3	7	-	-	-	-	-	10	6
Platymantis papuensis	2	-	-	6	6	-	-	14	7
Phrynomantis fusca	2	-	-	-	-	-	1	3	3
Rana modesta	9	-	-	1	-	-	-	10	7
Rana sp.*	12	-	-	-	-	-	-	12	12
Totals	35	7	11	7	6	1	1	68	53

*   Biologically undescribed species

8.2.1 Poro-poro

Poro also means 'foolish, silly' and although cognate with **poro-poro** it is unclear which derives from which. This is the only Nuaulu frog category consistently applied to one species only, namely *Litoria infrafrenata* (plate 13). The morphological and behavioural distinctiveness of this la ;e green tree frog, together with its commonness around villages makes this

understandable. The name is described by Nuaulu informants as onomatopoeic and given the call of this species, this seems highly plausible.

The reproductive biology of *Litoria infrafrenata* is well-understood by
the Nuaulu; that it lays large numbers of eggs in still water. These are
likened to sago jelly or porridge (**sona**), and this is no doubt connected with
the belief that **poro-poro** had its origin in the sago palm (*Metroxylon sagu*).
The palm features prominently in Nuaulu myths of origin of other life-forms,
including (notably) the Dutch.

PLATE 13: Two specimens of the tree-frog **poro-poro** (*Litoria infrafrenata*)
caught in banana plant in Rohua following rain: 14 August 1973 (neg. 73-4-
12).

8.2.2 notu

The term **notu** also means 'fart' (**notute** = fart (n. sing.), **inotu** = to fart,
to emit wind and noise from the anus). This association is a frequent source
of scatological humour, although it is uncertain which meaning is primary.
Poro-poro is occasionally used to refer to a particular kind of fart, but this
usage is almost certainly derived from its classificatory association with
notu.

The form most commonly referred to by this term was *Rana* sp., although it was found to be used to identify the following in order of decreasing frequency: *Rana modesta* (7), *Litoria* sp. (6), *Litoria amboinensis* (5.5), *Phrynomantis fusca* (3), *Litoria infrafrenata* and *Platymantis papuensis* (one each). Part of the explanation of this situation is undoubtedly to be found in the use of **notu** as a generic term for frogs (**notu kere, notu teteye,** etc.). It is clear that this is the sense in which respondents included the one specimen of *Litoria infrafrenata* which is otherwise consistently identified as **poro-poro.** This may also have been the case with *Platymantis papuensis, Phrynomantis fusca,* and *Litoria amboinensis,* which respondents also allotted to other categories. *Phrynomantis fusca* is encountered infrequently in the Nuaulu area and seems to have been allocated to the category **notu** in its generic sense. Both this species and *Platymantis papuensis* are also significant in that they are the only Seramese frogs known not to possess a tadpole stage; instead laying their eggs in damp places on land. However, limited knowledge of frog taxa among most informants suggests that it is by no means certain that they recognise folk categories or differentiate between separate species within the category **notu,** even if morphological and behavioural variation is acknowledged. Many informants were unable to attach more specific names in the context in which specimens were examined (usually in the village), and the fact that most of the specimens were collected by Menzies in the Jala river area and not by the Nuaulu themselves may well have contributed to their inability to differentiate further.

Rana sp. and *Rana modesta* are morphologically similar and the absence of lexemic differentiation is not surprising in the overall context of Nuaulu Anuran classification.

8.2.3 notu anae

Seven *Litoria* sp. of the *bicolor* group, very small green creatures, were described as **notu anae (anae** = 'child', 'young'), suggesting that although they were differentiated morphologically from other **notu** were not accorded a completely separate classificatory status, through either ignorance or disinterest. In fact, informants maintained that **notu anae** were subsequently transformed into **notu.** Komisi stated that **notu anae, notu** (in its terminal sense) and **kako** were all derived from the spawn of **notu** developmentally [c.f. Bulmer and Menzies, 1972-3a: 101-4].

8.2.4 kere

This term was explicitly stated by several informants to be onomatopoeic, the call being rendered as 'kere, kere, kere...'. Sometimes called **notu kere,** the term was applied most frequently to *Platymantis papuensis* although once to *Rana modesta.* In no case was **kere** applied to a frog for which no other term was also forthcoming. In the case of six specimens informants were equally divided as to whether *P. papuensis* was **kere** or **teteye.** In one case a specimen of *Rana modesta* was termed **kere** by one informant out of four, the other three settling for **notu.** One informant commented that **kere** was distinguished from **notu** on the morphological grounds that the former had a patterned skin (**unte nikate**), while the latter was black. This suggests that it is generally ascribed to *Rana modesta* rather than *Platymantis papuensis,* but the evidence is quite unclear. On all accounts **kere** was regarded by the Nuaulu as being very similar to **notu** (in its specific sense), although two older informants (Komisi and Sauute) did regard it as a type of **inararai.**

8.2.5 teteye

Probably an onomatope. This term was applied to *Platymantis* and no other species, although in no case was **teteye** applied to a frog for which another term was not forthcoming. In all cases the other term applied was **kere,** and in all cases while **kere** was the response given to dead and preserved specimens, **teteye** was that elicited from listening to tapes of calls of some of the specimens made before capture. This suggests that there is one classification based on morphology and another on call, although there is no evidence that the equation **kere = teteye** is recognised and the evidence of one informant that it is not. Indeed, one informant classified **teteye** with **kauke** (crickets and grass-hoppers) in a card sorting test on the basis of similar calls and the fact that both inhabit the underbush and rubbish heaps.

According to Komisi **teteye inae** (inae = 'mother') is a small frog, not much bigger than a thumbnail. I have no evidence as to whether this is simply a small **teteye** or a different species altogether.

8.2.6 inararai

Only one response elicited this term and then for *Litoria amboinensis,* for which all other responses were **notu.** I have already suggested that informants' descriptions of *L. amboinensis* as **notu** involve the use of the term generically and perhaps on account of a poor knowledge of the anatomy of *Litoria.* That **inararai** is generally used for *L. amboinensis* is confirmed through the coincidence of names elicited independently for the call and on the basis of morphology. In fact, call seems particularly important in identifying *L. amboinensis,* which may well explain the allocation of a live but noiseless specimen to the category **notu.** However, at least two informants (Komisi and Inane) said they were able to recognise **inararai** on the basis of its possession of a prominent tail remnant, in which case observations were presumably based on young, incompletely metamorphosed, specimens. **Inararai** is generally described as a type of **poro-poro** on account of the fact that, like **poro-poro,** it is said to have originated from **notu.** Two types of **inararai** are usually said to exist: **inararai marae** (**marae** = 'bluegreen'), a small green tree frog, and **inararai msinae** (**msinae** = 'red'), a reddish and heavily mottled form with a yellowish ventral surface becoming deeper on the throat. There is some evidence that these may represent distinct colour morphs.

8.2.7 kako (kakoi)

The term is onomatopoeic and sometimes rendered **kako-kako.** It is occasionally confused with **kako nione** (**nione** = coconut), which is how the vocalisation of **nopa hanaie** (chapter 6.2.7.5) is described. Only one specimen of *Phrynomantis fusca* was obtained from near Rohua to which informants attached this label, although it was very much in evidence vocally. **Kako** was described to Ellen as a distinct type, smaller than **notu,** from which it is said to develop. Like **notu,** it only lays a few eggs at a time.

8.3 Social uses of frogs

Poro-poro is the only frog reported as being eaten by the Nuaulu, although there are no prohibitions on consuming other types. Although it is probably used as a general famine food it is commonly caught by children in the context of play activities and roasted whole in an open fire. With such an abundance of potential sources of animal protein (particularly in the form of 'big' game animals) only the larger of the small game species are regarded as being worthwhile collecting, unless they have other qualities which make them desirable. At least one Nuaulu (Saite Somori) used **poro-poro** to catch young eels, as an English countryman might use a ferret to catch rabbits.

No ritual associations have been reported for frogs other than the existence of a **wate poro-poro.** This is a scare charm or taboo sign used by the clan Matoke to threaten actual or potential thieves. Violation is said to give rise to stomach ache and itching in the victim.

8.4 General remarks on the classificatory structure of categories for frogs

Although no single term is used exclusively for frogs as a whole, it is clear from observation, card sorting tests and interviews that they are seen as a distinctive 'natural' group. The term **notu** is usually used to refer to all unspecified frogs, although this term contrasts at several different degrees of inclusiveness:

1) **notu:** all non-frogs

2) **notu : poro-poro** (i.e. tree frogs)

3) **notu : poro-poro, teteye, inararai, kere**

4) **notu : kako, notu anae**

At the second degree of inclusiveness **notu** is used for all ground and bush-dwelling frogs, in contrast to tree frogs (**poro-poro**). Alternatively, this can be seen as a distinction between frogs of the river and forest (sometimes labelled **notu waene; waene** = 'river', 'freshwater') and those of the village and gardens, although it is recognised that this is not an accurate classification of natural kinds in terms of habitat. The second degree of inclusiveness of **notu** includes all terminal categories with the exception of **poro-poro** and **inararai,** which are usually grouped together under the separate generic **poro-poro.**

The third contrast distinguishes all other types from **notu** (in its terminal sense), **kako** and **notu anae.** We have already noted that these last two are held to develop from **notu** spawn, and this presumably is part of the logic grouping them together.

These contrasts give us the taxonomic structure set out in figure 12. Such a taxonomy seems to explain the various contrasts made in a way most consistent with Nuaulu knowledge of anuran biology. It is not suggested that it is a consistently employed mental construct; rather it is an analytical aid to interpretation. A different means of modelling is presented in figure 13. The advantage of the Venn diagram is that it plays down the cognitive centrality of hierarchy and contrast, emphasising the fuzziness of the relationships.

It is clear from what has been said that behaviour (especially vocalisation) is an important means of distinguishing between different types of frog. In 1975 Menzies made a series of recordings of frog vocalisations. Two informants were able to agree on the identification of three types of frog on the basis of their vocalisations, which matched exactly the names elicited on

FIGURE 12 Nuaulu classification of frogs arranged as a taxonomy. The chart incorporates the *maximum* number of classificatory levels elicitable from informants. In all cases the condition of transitivity is understood.

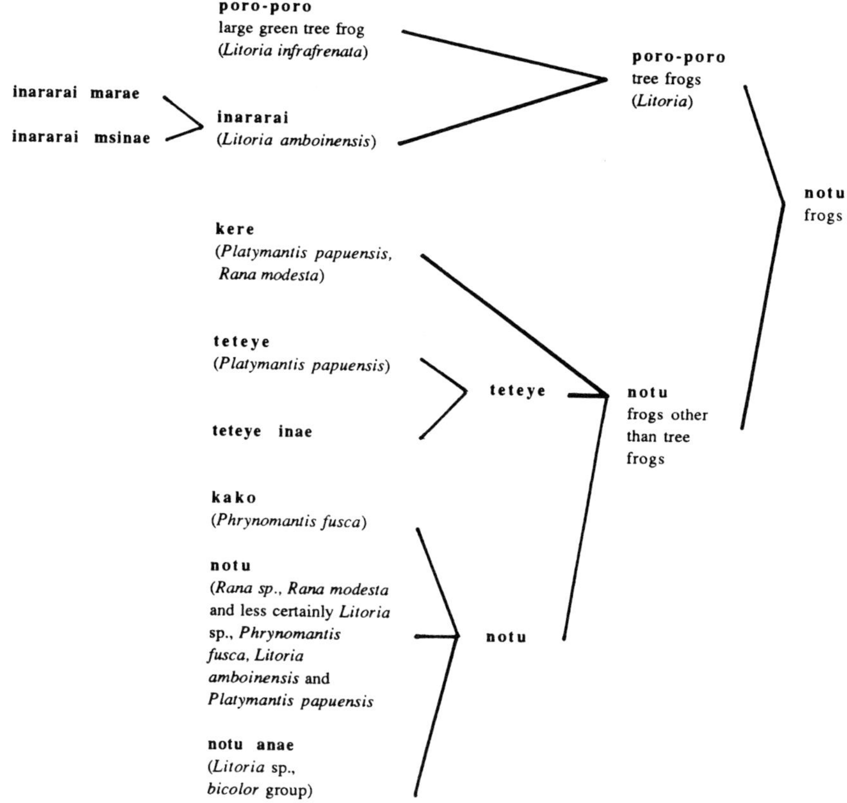

FIGURE 13 Nuaulu frog classification arranged as a Venn diagram. In the figure La, Lb, Li, Pf, Pp, Rm and R? refer to Linnaean nomenclature spelled out in figure 8.1.

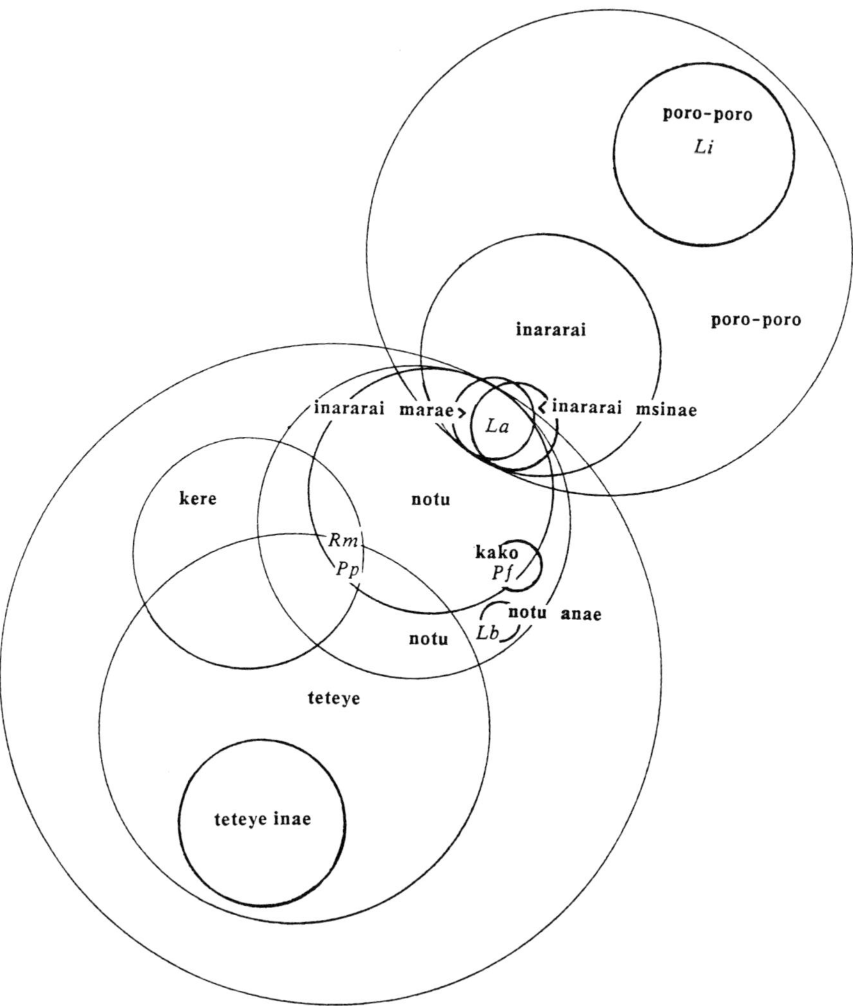

independent examination of the live specimens. The case of **teteye** is particularly interesting in this respect. On one occasion Komisi described some specimens of *Litoria amboinensis* as **notu waene,** that is 'river', 'stream' or 'freshwater' **notu.** He said that three kinds could be recognized on the basis of the sound they produced:

1) those calling from holes in stones (nocturnal),

2) those calling from under the water (diurnal), and

3) those calling from the dry land - **inararai.**

The only positive identification here is the third as *Litoria amboinensis.* It is difficult to know to what extent these represent a knowledge of actual frog behaviour and to what extent they are simply colourful ways of expressing particular vocalisations.

8.5 Consistency in the application of frog labels

I think it is clear that there is a discernible structure to Nuaulu classification of frogs and that it is not a wilful distortion to represent this for didactic purposes as a taxonomy. On the other hand, despite the sophistication of some of the discriminations employed, the evidence presented here indicates that boundaries between categories are often operationally 'fuzzy'. Knowledge of frogs is poor and identification inconsistent, compared with other vertebrate groups known to the Nuaulu and compared to other societies for which there are data [Bulmer and Tyler, 1968].

In July 1975 Menzies and Ellen visited the non-Nuaulu mountain village of Piliana, at an altitude of 700 m above Japutih on Teluti Bay. Here a small collection of frogs was made and some information on indigenous terms gathered (for a list see Ellen et al, 1976b: table 3, p. 136). Although the period at Piliana was brief (a few days only) and the material gathered entirely through the medium of AM, it was clear that Pilianan knowledge of anuran biology was more extensive than that of the Nuaulu and that terms were applied to biological species with a much greater consistency, as well as there being a much closer correspondence between phylogenetic and local categories. This appears to be due to a combination of the greater diversity and population of frogs in this upland region and more restricted sources of animal protein. The latter compels inhabitants to take a greater interest in minor food resources. Only three out of seven species were not thought to be appropriate foods, two because they were simply too small and one because there were no local or regional traditions prescribing this as a suitable food source.

Nuaulu are not compelled to take such minor protein sources as frogs over seriously, as other more reliable and productive ones are available. Lack of nutritional significance is probably the most important single reason explaining the restricted anuran inventory and inconsistency in the application of terms. However, locality is also important. Although the smaller frogs undoubtedly do occur in the inner zone of the Nuaulu extractive environment (the area bounded by the most distant gardens), some species are only found in abundance at more distant localities, determined by altitude (*Rana* spp.) and extensive swamp forest (*Phrynomantis fusca*). High altitude areas are rarely visited, even on hunting expeditions and for the collection of *Agathis* resin. Areas of swamp forest, however, are commonly visited for the purpose of collecting sago flour and their fauna is well-understood. However, that the nomenclature is as extensive as it is, gives some reason to suspect that it evolved during the period the Nuaulu were living in the highlands. Migration, resettlement and economic change have combined to alter the cultural significance of frogs, with the result that while the names remain the experience necessary to employ them constantly, consistently, and perhaps also accurately, is lacking.

CHAPTER NINE

FISHES AND MARINE MAMMALS

9.1 Fishes and marine mammals of the central Moluccas

Wallace [Wallace, 1962 (1869)] noted that the fishes of the central Moluccas were perhaps unrivalled for variety and beauty. More than 780 certain freshwater and marine species are known from the area, although the number of species encountered during the course of fieldwork in the Nuaulu area was extremely small. For the most part, this reflects Nuaulu lack of knowledge of maritime fauna, and only secondarily relative geographical distribution. Few specimens were collected, and identifications are based for the most part on photographs, drawings and written descriptions.

A checklist of some of the major fish reported for south central Seram, including those forms with which the Nuaulu are familiar, is presented in table 15.

9.2 Nuaulu categories applied to fish

9.2.1 woku

AM = 'ikan yu' or 'kaluju'; Ind. 'ikan hiu'. Applied to all non-bony fishes other than *PRISTIFORMES* encountered by the Nuaulu.

9.2.1.1 woku ikae

Ikae = 'fish'. Applied to all typical sharks. These are occasionally seen off the coast of south Seram, but never sought after and actively feared.

9.2.1.2 woku sanihuhue

The meaning of this term is unclear. Refers to hammerhead sharks.

9.2.1.3 woku karakaiya

Lit. 'saw shark'. In 1969 Sauute Nepane-tomoien found a sawfish trapped in his net off Rohua. This was an unusual incident. The fish was eaten and the 'saw' preserved as a trophy.

9.2.1.4 woku sonu

Sonu is the term applied to needlefish (9.2.13). **Woku sonu** is not a shark at all, but swordfish of the perciform family Xiphiidae. None were observed in the Nuaulu area during fieldwork.

9.2.2 (ika) hari

Glossed in AM as 'ikan pari'. Applied to sting-rays, which are sometimes encountered when fishing or travelling along the coast in an outrigger canoe. In 1970 I had to treat a youth who had been badly stung by a ray.

9.2.3 ama-ama

Lit. 'father-father'; applied to a small anchovy *Thrissina baelama*, glossed in AM as 'ikan lumpa' or 'ikan lompar', and compared with 'ikan nasi'.

9.2.4 ika siori

Cognate with AM 'ikan teri', sometimes also known as **puri.** This fish has a black dorsal surface, silver ventral surface, orange markings on the face and a golden tail. The term is applied to anchovies of the genus *Stolephorus.*

9.2.5 ika mate

Glossed in AM as 'maki, make'; applied to *Clupea schrammi, C. stereolepsis* and other herrings.

9.2.5.1 mate inae

Inae = 'mother'; applied to *Clupea longiceps* and *Sardinella fimbriata.*

9.2.5.2 mate seran

Seran is often treated as an acceptable variant of 'Seram', though in this case the reference is probably to that part of eastern Seram originally designated by the term. For the Nuaulu Seram as a whole is **nusa ina,** 'mother island'. Applied to *Spratelloides delicatulus* and other sprats.

9.2.6 (ika) petie

Peti is the root of a number of words meaning to tremble, vibrate, beat and knock. Applied to freshwater pikes including *Ophiocephalus striatus* and almost certainly *O. micropelties.* AM gloss = 'ikan gabus'.

TABLE 15 Checklist of fishes and marine mammals recorded in and around the Nuaulu region of south central Seram, 1970-75.

Species	Ecological zones					Nuaulu gloss
	1	2	3	4	5	
LAMNIFORMES - sharks)	-	-	-	+	+	**woku**
typical sharks	-	-	-	+	+	**woku ikae**
Sphyrnidae - hammerhead sharks						
Sphyrna sp.	-	-	-	+	+	**woku sanihuhue**
MYLIOBATIFORMES						
Dasyatidae - stingrays						
Dasyatis kuhli	-	-	-	-	+	**hari**
PRISTIFORMES						
Pristidae - sawfish						
Pristiopsis sp.	-	-	-	-	+	**woku karakaiya**
CLUPEIFORMES						
Engraulidae - anchovies						
Thrissina baelama	-	-	-	+	-	**ama-ama**
Stolephorus spp.	-	-	-	+	-	**ika siori**
Dussumieriidae						
Spratelloides delicatulus	-	-	-	+	-	**ika mate seran**
Clupeidae - herrings						
Clupea schrammi	-	-	-	+	-	**mate**
Clupea stereolepsis	-	-	-	+	-	**mate**
Clupea longiceps	-	-	-	+	-	**mate inae**
Sardinella fimbriata						
Esocidae - pikes						
Ophiocephalus striatus	+	-	-	-	-	**petie**
Ophiocephalus micropelties						
	+	-	-	-	-	**petie wae nosite**
	+	-	-	-	-	**peti kunie**
CYPRINIFORMES						
Cyprinidae - carps	+	-	-	-	-	**ika timanne**

SILUROIDIFORMES
Plotosidae - catfish eels
Paraplotosus albilabris
Plotosus anguillaris
Various freshwater catfish

					wotu-wotu
+	-	-	-	-	namue
+	-	-	-	-	nam hosu
+	-	-	-	-	nam wanae

ANGUILLIFORMES - eels and morays
Anguillidae - freshwater eels
Anguilla bicolor pacifica
Anguilla celebensis
Anguilla marmorata

+	-	-	-	-	awane
+	-	-	-	-	awane tunne
+	+	-	-	-	ika koa totue

Muraenidae - moray eels
Siderea picta
Muraena spp.
and other spp.

-	-	+	+	-	yapato
-	-	+	+	-	

BELONIFORMES
Belonidae - needlefish
Strongylura
Tylosurus
Hemiramphidae - halfbeaks, garfish
Hemiramphus marginatus
Zenarchopterus - river garfish
Various saltwater genera
Exocoetidae - flying fish
Cypselurus spp.

-	-	-	+	+	sonu
-	-	-	+	+	sonu
-	-	-	+	+	wanu
+	+	-	-	-	wanu
-	-	-	+	-	wanu
-	-	-	+	-	(ika) keuro

GADIFORMES
Bregmaceros macclellandi

PLEURONECTIFORMES - flatfish
Unidentified flatfish

+	-	-	-	-	ika rina

SYNGNATHIFORMES
Fistulariidae - cornet fish
Fistularia eptimba
Syngnathidae - pipefish, sea horses
Hippocampus polytaenia

-	-	+	+	-	peskada

MUGILIFORMES

Sphyraenidae - barracudas

Sphyraenella chrysotaenia	-	-	-	+	+	taniri

Mugilidae

Mugil troscheli	+	-	-	-	-	hoi waene
	-	-	-	+	+	hoi nuae

Atherinidae - silversides

Pranesus duodecimalis

Melanotaenidae - rainbowfish	+	-	-	-	-	hoi manikate

PERCIFORMES

Scombridae - mackerels

Rastrelliger spp.	-	-	-	+	-	lema

Scomberomoridae

Scomberomorus						taniri

Thunnidae - tunas and albacores

Auxis thazard	-	-	-	+	+	komu

Thynnus albacares

Katsuwonus pelamis

Kishinoella tonggol	-	-	-	+	-	pia

Euthynnus

Istiophoridae - sailfish and
related forms

Istiophorus orientalis	-	-	-	+	-	ika nane
Xiphiidae - swordfish	-	-	-	-	+	woku sonu

Trichiuridae - cutlass fish

Trichiurus haumela	-	-	-	+	-	tumane-maine

Carangidae - jacks, scads and
pompanos

Selaroides leptolepsis	-	-	-	+	-	ika hutua onate
Caranx spp.	-	-	-	+	-	heti
Alepes mate	-	-	+	+	-	heti

prob. *Alectis* sp.

Decapterus macrosoma	-	-	-	+	-	momar

Lutjanidae - snappers

Lutjanus fulviflamma	-	-	-	+	-	ika ka
Lutjanus rufolineatus	-	-	-	+	-	ikae msinae, ika iloru
Lutjanus sanguineus	-	-	-	+	-	ika mala

Caesiodidae.

Caesio spp.	-	-	-	+	-	ika uri hatae

Theraponidae - tiger fish						
Therapon	-	+	-	+	-	**ika kutulauno**
Mullidae						
Unidentified goatfish	-	-	-	+	-	**ikae uri hatae**
Pempheridae						
Pempheris sp	+	-	-	-	-	**eta-eta**
Chaetodontidae - butterfly fish, coral fish	-	-	+	-	-	**ikae kori-kori**
Heniochus acuminatus	-	-	+	-	-	**ikae hanu totue**
Coridae - wrasses, rainbowfish						
Cheilinus spp.	-	-	+	-	-	**punu**
Stethojulis kalosoma	-	-	-	+	-	**sunu moti**
unident. wrass	-	-	+	-	-	**ika tuyo hutan**
Scaridae - parrotfishes						
Callyodon or *Scarus*	-	-	-	+	+	**ika nakatua**
Blenniidae - blennies	-	-	+	-	-	**ika lasiato**
Acanthuridae - surgeon fish and related forms						
Acanthurus lineatus	-	-	+	-	-	**ika hanu totue**
Gobiidae - gobies						
Paragobiodon echinocephala	-	-	+	-	-	**ika lasiato**
Periophthalmidae - mudskippers						
Periophthalmus vulgaris	-	+	-	-	-	**lasiato**
Scorpaenidae - scorpion fish and rock fish						
Parascorpaena maculipinnis	-	-	+	-	-	**ika sinatane**
Percidae -perches						
Anabas testudineus - climbing perch	-	+	-	-	-	**ika sa sahune**
Lactaridae						
Lactarius lactarius	-	-	-	+	-	**lema**
TETRODONTIFORMES						
Diodontidae - porcupinefish *Diodon*	-	-	-	+	-	**ika tunene**

	1	2	3	4	5	
Balistidae - trigger fish and file fish						
Balistes spp.	-	-	+	-	-	**ikae uro, ikura**
Monocanthidae						
Tetraodontidae						
	-	-	-	+	+	**sunu**
Unidentified TELEOSTOMI (bony fish)	+	-	-	-	-	**wanapoi**
	-	-	+	-	-	**ika moti atue**
	+	-	-	-	-	**ika ai hata kopue**
	+	-	-	-	-	**monosia**
Aquatic mammals						
CETACEA						
whales	-	-	-	-	+	**ika baus**
dolphins and porpoises	-	-	-	-	+	**ika hahu**
SIRENIA						
Dugong dugon	-	-	-	+	-	**mata luyun**

Key: Zone 1 = freshwater; zone 2 = brackish water, mangrove swamp; zone 3 = littoral, reefs and rock pools; zone 4 = offshore; zone 5 = deep sea.

9.2.6.1 peti(e) wae nosite

Lit. '**petie** of the river bottom', referring to the habit of this black and white freshwater fish of laying in the sand of river bottoms.

9.2.6.2 peti kunie

Lit. 'tumeric **petie**'; tumeric, *Fibraurea chloroleuca*, being a woody climber used widely as a source of yellow dye. The dorsal surface of this freshwater fish is red and black with yellow fin extremities; the ventral surface is yellow with reddish fin extremities.

9.2.6.3 peti rihu-rihue

Rihu = 'fly(ing)'.

9.2.7 ika timanne

Glossed in AM as 'ikan mas' ('gold fish'); applied to various species of freshwater carp.

9.2.8 wotu-wotu

Glossed in AM as 'ikan sembilan' ('nine fish'), this species is regarded as dangerous; applied to *Paraplotosus albilabris* and *Plotosus albilabris*.

9.2.9 namue

Term applied to a freshwater fish with a broad mouth and depressed body form, mostly catfish. Commonly fished.

9.2.9.1 nam hosu

Meaning of qualifier unknown. Unidentified black catfish with small orange spots sparsely scattered over surface.

9.2.9.2 nam wanae

Prob. 'child of the **namue**. Unidentified freshwater catfish.

9.2.10 awane

Awane, wane = 'lianas, trailers, string'. AM gloss = 'muria air' and applied to freshwater eels (Anguillidae), including *Anguilla bicolor pacifica*. All **awane** are totemic for the clan Pia. Five types are recognised, though the fifth is optional.

9.2.10.1 awane metene

Metene = 'black'. Also known as **awane (muna) nahu, awane unu panane** and **awane mnanahu (mata putie)**. **Nahue** refers to palm fibre, while **muna** is a verb meaning 'to adversely affect in some way'; **unu panane** means 'the top of the head'. **Awane muna nahu** is sometimes distinguished from other **awane** on the grounds of its much smaller size.

9.2.10.2 awane msinae

Msinae = 'red'. This is also known as **awane tunne, tunne** meaning 'very' as in 'very best'. The term presumably refers to the attractive markings of this animal, *Anguilla marmorata*.

9.2.10.3 awane putie

Putie = 'white', for which there are no recorded synonyms.

9.2.10.4 awane hunanene

Hunane = 'moon' + **-ne** (third person singular possessive); therefore lit. 'eel of the moon'. Reference unclear.

9.2.11 ika koa totue

Lit. 'pandanus leaf fish', referring to the appearance of this attenuated and extremely laterally compressed form of fresh and salt water. A term apparently applied to Anguillid elvers, but may also extend to other eels.

9.2.12 yapato

Yapato is possibly a recent (re-introduced?) loan word, in view of the fact that the Nuaulu have only been located on the coast for a historically short period. Applies to saltwater eels, probably moray eels (Muraenidae), including the locally present *Siderea picta*. **Yapato** (which is also often glossed with AM 'ikan muria') may sometimes be used to refer to certain seasnakes which resembles eels quite closely. The Nuaulu distinguish two terminal categories:

9.2.12.1 yapato putie

Putie = 'white'. Whitish or yellowish speckled dorsal surface.

9.2.12.1 yapato msinae

Msinae = 'red'. Reddish speckled dorsal surface.

9.2.13 (ika) sonu

Sonu may be an archaic word for 'sago'. Often glossed in AM as 'ikan sagu', the association between it and a sago palm is clear from an origin myth in which the Dutch are reputed to have first emerged from the trunk of a sago palm (a common Nuaulu mythic theme) and then transported to Holland with the assistance of a friendly **ika sonu**. The existence of such a story, and the apparent fact that the term is not a recent loanword, suggests that familiarity with the fish is longstanding.

Applied to needlefish of the genera *Strongylura* and *Tylosurus,* which are caught using the web of the round-bodied spider **ka(h)uneke wala-wala sonu** (chapter 13.2.12). The head of a caught **ika sonu** is never discarded, but placed between the slats of the wall of a house near the seaward entrance (**mitanunue nuae**). It is regarded as a trophy well worth displaying as well as acting as a charm to bring good fortune in subsequent fishing (see BM specimen As. 1. 312, and plate 14).

PLATE 14: Dried head of needlefish (**sonu**: *Strongylura* or *Tylosurus*) displayed as trophy on seaward house entrance. L. 13.5 cm. (As 1.312). Photograph reproduced by courtesy of the British Museum.

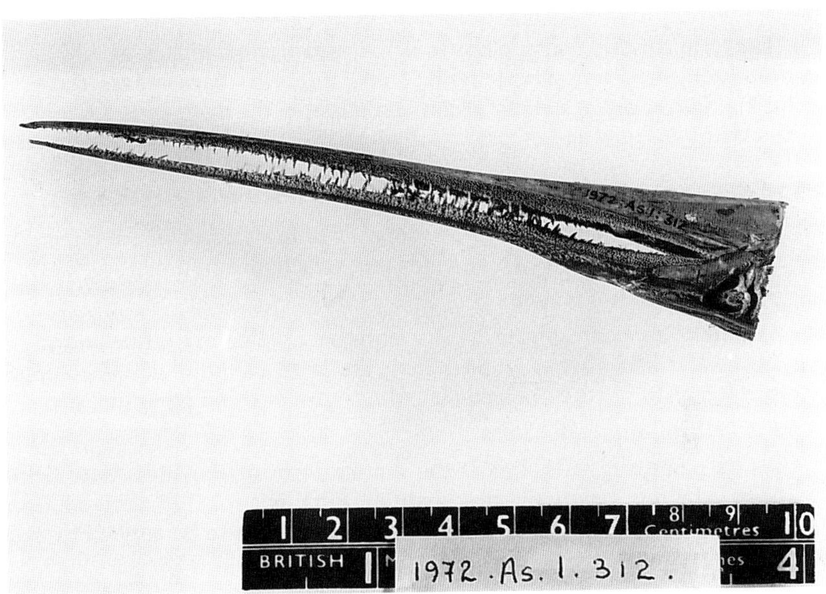

9.2.14 wanu

Glossed in AM as 'julung-julung'; applied to the halfbeak *Hemiramphus marginatus* and other Hemiramphid genera.

9.2.15 ika keuro

Flying fishes, including *Cypselurus nigripennis*, known in AM as 'tuing-tuing', No particular focal form identified.

9.2.16 ika rina

Term applied to an unidentified species of freshwater flatfish. It is about 11 cm long, with a black dorsal surface and white ventral surface.

9.2.17 peskada

From Portuguese 'pes' or 'pescada' (fish, kind of fish) modified by a perceived identity between 'cada' and AM 'kuda', meaning horse; lit. 'horse fish'. There is no syllable final /s/ in Nuaulu, nor an /sk/ consonant cluster

(R.B.). Particular focal species of true sea horses, but probably extended to include pipefish. Applied to specimens of *Hippocampus*.

9.2.18 taniri

The term for this fish appears to be derived from cognates of AM 'ikan tenggiri (tengguri, tengjuri, tengjiri)', and refers to barracudas, certainly including *Sphyraenella chrysotaenia*, and possibly the mackerels *Scomberomorus*, *Auxis thazard* and others.

9.2.19 (ika) hoi

Hoi may be an adjectival contraction of **hoine** ('eczema'), referring to the appearance of this fish. Three kinds recognised, one optional.

9.2.19.1 (ika) hoi waene

Waene = 'river, freshwater'. The terms are applied to a white and silver freshwater fish, about 13-15 cm long, and with a slightly darkened upper surface. AM gloss = 'bulana air': *Mugil troscheli* and other mullets.

9.2.19.2 hoi nuae, hoi saha

As its name suggests, this is the marine form of **ika hoi**; **nuae** being 'sea'. In the synonym, the qualifier **saha** is probably derived from **msaha**, meaning 'newly married husband'. Seems to be applied to various saltwater mullets.

9.2.19.3 hoi manikate, ika manikate

Man ikate means 'bird pattern', referring to the markings of this silver freshwater fish with a dark dorsal surface. Possibly a rainbow fish of the family Melanotaenidae; glossed in AM by the term 'loreng-loreng'.

9.2.20 lema

Cognate with AM 'laima' or 'lemah'. Applied to one of the most commonly caught types in offshore fishing along the Seramese coast, the mackerel *Rastrelliger neglectus*. Those specimens caught by the Nuaulu were generally confined to the young. AM 'lemah' is also known to be applied to whitefish, *Lactarius lactarius*.

9.2.21 komu

Applied to the tuna *Auxis thazard*.

9.2.22 ika pia

Pia refers to the river of the same name 1 km east of the village of Bunara (see figure 3). Glossed in AM as 'cakalan': tuna, focally *Kishinoella tonggol*.

9.2.23 ika nane

Lit. 'sail fish', making it semantically equivalent to the AM gloss 'ikan layar', indicating *Trichiurus savula*; but perhaps also including the spectacular pacific sailfish *Istiophorus orientalis*.

9.2.24 tumane-maine

Lit. 'wriggling axe'; AM gloss = 'ikan mencadu' : 'axe fish'. Possibly an allusion to *Trichiurus haumela* or near, also known in AM as 'parang parang'.

9.2.25 ika hutua onate

Lit. 'large intestine fish'; in AM 'ikan ila (aela). Probably *Selaroides leptolepsis*.

9.2.26 ika heti

The term is glossed in AM as 'ikan bubara, bobara': various trevally of the genus *Caranx* (AM 'kawalinja'), probably including *C. armatus* and *C. melampygus*; also possibly *Alepes mate* and *Alectis*.

9.2.27 momar

Cognate with identical AM term; applied to *Decapterus macrosoma*.

9.2.28 ika ka

One specimen of this marine form was identified as the snapper *Lutjanus fulviflamma*.

9.2.29 ika iloru, ika msinae

Refers to a non-Nuaulu village of that name, possibly Ileri near Nueli-tetu (see figure 3) which no longer exists but which is marked on some old maps: Topographische Inrichting 1918, *Schetskaart van Ceram* (Scale 1 : 100, 000), Batavia; and others. In the synonym, **msinae** = 'red'. Glossed in AM variously as 'geropa' and 'ikan merah (gurara)'; applied to the red snapper *Lutjanus rufolineatus*.

9.2.30 ika mala

Applied to *Lutjanus sanguineus*.

9.2.31 ika uri hatae

Lit. 'banana branch fish'. This saltwater fish is dark brown with red bands, and with black spots surrounded by a light grey ring behind the eyes. Glossed in AM as 'pisang-pisang' and 'ikan salmonet'; considered to be related to AM 'lalosi'. Probably *Caesio pisang*, extended to *C. erythrogaster* and *C. chrysozona*.

9.2.32 eta-eta

Term applied to a small silver freshwater fish with a laterally compressed body form, possibly *Pempheris*.

9.2.33 ika kutulauno, ika utu-utu

The name is possibly an allusion to rough (**kutu-kutue**) waves (**kolau**, pl.); **utu** = louse (10.2.8). A term applied to two closely related types of fish which are not terminologically distinguished by the Nuaulu themselves. A spotted form is glossed in AM as 'kerong-kerong', and a non-spotted form as 'ikan lasi'. Both are known to be eaten by storks. The term is applied to *Sphyraenella chrysotaenia*, but may also be extended to include sea basses such as *Epinephelus caeruleopunctatus* and *E. fuscoguttatus*.

9.2.34 ika kori-kori

Lit. 'butterfly fish', which equates semantically with the attributed AM gloss, 'ikan kopu-kopu'. Perhaps unsurprisingly, this is a label applied fairly generally to a wide range of saltwater butterfly and coral fishes, and perhaps also other brightly-coloured fish of the reefs and rock pools.

9.2.35 ika hanu totue

Hanu is a type of *Hibiscus* shrub, so literally this is 'hibiscus leaf fish'. L = about 20-24 cm; possibly an Acanthurid coral fish (nr. *Acanthurus lineatus*), but also applied to the butterfly fish described in AM as 'ikan bendera', flag fish (e.g. *Chaetodon lunula* and *Heniochus* spp.).

9.2.36 (ika) sunu, sunu moti

This term describes fish found in rock pools at low tide (**moti**), and glossed in AM as 'tatu'. Applied to rainbow fish of the genera *Cheilinus* and *Stethojulis*.

9.2.37 ika tuyo hutan

Lit. 'seven forests fish'; cognate with AM, from which it is apparently derived. A small fish of the seashore and rockpools with a dorsal surface of dark brown, black and orange. The ventral surface of the head is yellow, and the dorsal surface markings are dark crimson and canary yellow. The ventral part of the head is bounded by a tourquoise line. The specimens so-labelled were all juveniles: an otherwise unidentified wrass. Secondary totem for clan Numanaita.

9.2.38 ika nakatua

Lit. 'cockatoo fish', being an allusion to the 'beak' of this large parrot fish: *Callyodon* or *Scarus*. 'Ikan kakatua' in AM.

9.2.39 ika lasiato

Term cognate with Sepa 'lasiato', from which it is almost certainly derived. Meaning unclear. Focally, probably mudskippers (*Periophthalmus vulgaris*), but also applied to some local gobies and blennies.

9.2.40 ika sinatane

Sinatane are nettles, including *Laportea stimulans* and *Procris frutescens;* apparently an allusion to the sting this fish is capable of giving (but see also **kori-kori sinatane;** chapter 10.2.34.9). Applied to a scorpion fish of the rockpools, and glossed by the AM terms 'ikan bia' and 'ikan suangi'. May also refer to a separate freshwater species. **Ikae sinatane** is mashed with red coconut to give a hunting potion for the clan Matoke.

9.2.41 ika sa sahune

Lit. 'climbs **sa(h)une**', the pandan *Pandanus conoideus.* Probably refers to the saltwater climbing perch, *Anabas testudineus,* which is known to be able to live outside water for lengthy periods with the aid of a labyrinth organ. **Sa sahune** is applied to pond skaters (10.2.15).

9.2.42 ika tunene

Tunene is the durian, *Durio zibethinus.* This name appears to be an allusion to the prickly skin of porcupine fish, though the precise species to which it is applied are unknown.

9.2.43 sunu

Includes the pufferfish *Tetraodon*, and possibly also triggerfish and filefish (Monocanthidae). Notable for toxicity and venomous fin spines; flesh considered poisonous. Note that this is not the same category as 9.2.36.

9.2.44 ika uro, ukura

Meaning unknown; applied to a species of *Balistes*.

9.2.45 wanapoi

Unidentified large freshwater fish.

9.2.46 ika moti atue

Moti refers to rocks and coral areas exposed at low tide; **ika moti** is therefore, in a general sense, 'reef fish'. The meaning of **atue** in this context is unknown. The term is applied to fish of many species not distinguished terminologically which are found in rock pools at low tide.

9.2.47 ika ai hata kopue

Lit. 'rotting sago tree fish'. This freshwater fish is glossed in AM as 'gabus'.

9.2.48 (ika) monosia

Term identical in AM; a freshwater species.

9.2.49 (ika) pono

This fish, reported for both marine and freshwater localities, is glossed in AM as **ikan tanira.**

9.2.50 (ika) unone

Glossed in Sepa as 'iano meme wailo'; unknown freshwater species.

9.2.51 ika baus, baus baus

From AM 'ikan paus', paus fish', lit. 'pope fish'. A term applied to all whales seen, heard or heard of by the Nuaulu. I have one report of the term **arauwane** for 'whale', but its status is unclear. The most common whales sighted in Seramese waters appear to be the sperm whale, *Physeler catadon* and various baleen whales, *Balaenoptera* spp. Occasional strandings are known.

9.2.52 ika hahu

Lit. 'pig fish', semantically cognate with AM 'ikan babi' for the same; although it is also known as 'lumba-lumba'. The term is applied to dolphins, most of which are probably of the genus *Stenella*. However, the distinction between dolphins seems unlikely to be a clear one, not only for the Nuaulu. Indeed, the people of Lamalera on Lembata use the term 'temu' to include short-finned pilot whales (*Globicephala macrorhynus*), the grampus (*Grampus griseus*), the false killer whale (*Pseudorca crassideus*), as well as most dolphins [Hembree, 1980: 11-6].

9.2.53 (mata) luyun

Cognate with AM term meaning 'moon eye', and applied to the dugong, *Dugong dugon*. Few Nuaulu have ever seen this creature, which is rare in the waters of south central Seram. Sauute Nepane-tomoien caught a dugong in the fifties, and his son Sekanima caught a stray female off the Mon headland in January 1990. Nowadays dugong are best known from the shallow waters around the Seram Laut and Gorom archipelagos, where their teeth and ribs are still occasionally carved into objects such as smoking pipes. Dugong has a reputation for human-like behaviour and was formerly valued for the tusk-like incisors of the male which were made into ear-rings and obtained by the Nuaulu through trade. Its flesh is eaten by the Nuaulu when available, the teeth and 'tears' (the copious emission from the lachrymose gland) sold to Chinese traders, and the ribs and skull kept for magical purposes. Nuaulu consider it to be the ancestor of all Chinese.

9.3 The social and economic significance of fishes and marine mammals

Fish of all kinds constitute a sizable proportion of all animal protein consumed by the Nuaulu, around 25 percent consumed weight judging from the results of a four month dietary survey [Ellen 1993: chapter 6.3]. Most fish eaten by the Nuaulu are either of freshwater origin or bought from other fishing peoples along the coast. This includes virtually all tuna (**ika pia**) consumed by the Nuaulu. Freshwater fish are obtained through traditional methods employed in the mountain homelands: fish-trapping, mud-scouring, stunning with plant poisons, and using bow-and-arrow. Since coming to the coast, a few individuals have taken to marine fishing: night-time lamp fishing for squid, and the use of light surrounded by a net for catching bait fishes such as *Stolephorus, Caesio, Sardinella* and *Caranx*. A type of purse seine fishing ('giob', in AM) is also sometimes employed for *Caranx, Decapterus, Hemiramphus* and *Sphyraena*. Pole and line live-bait fishing takes place from the beach, and fishes stranded in rock pools are collected at low tide using basket traps.

There is virtually no ritual restriction on the consumption or procurement of particular species, and almost anything available will be eaten, although no attempt is made to catch the larger fish and marine mammals, which are only eaten when they are occasionally thrown up on beaches, get caught in nets or stray out of their natural habitat. This catholic attitude to the consumption of marine fish in particular, and the absence of strong traditions relating to marine species, reflects the fact that the Nuaulu have been a coastal people for less than 100 years.

TABLE 16 Some non-assimilated Sepa fish terms occasionally used by Nuaulu.

Species	Sepa	AM gloss
SYNGNATHIFORMES		
Fistulariidae - cornet fish		
Fistularia petimba	tuano	ikan trompet
PERCIFORMES		
Chirocentridae		
Chirocentrus dorab		
Trichiuridae - cutlass fish		
Trichiurus haumela	tahula	ikan parang-parang
Carangidae - jacks, scads and pompanos		
Selar boops	pulala/papalo	palala/papalo
Lutjanidae - snappers		
Caesio cuning	wau'uno	lalosi
Pomacentridae - damselfish		
Abudefduf	matenu	ikan papua

9.4 Some general features of Nuaulu knowledge and classification of fishes

The category **ikae** is approximately co-terminus with AM 'ikan'. It includes all marine mammals, but there is a fuzzy area of overlap with respect to eels, sea-snakes and perhaps freshwater snakes. Marine coelenterates and echinoderms, such as sea-stars, are closely linked to **ikae** by ecological association and the term is sometimes used to describe them. When appearing as the first element in a binomial, **ika** and **ikae** appear to be used more-or-less interchangeably, though some terms seem to display one more frequently than the other. In presenting the terms here I have made no attempt to systematize this usage and have usually given the form as elicited.

Nuaulu terminology for fish is very restricted when compared with the richness of the fauna of the inshore waters of Seram, and the terminologies of other coastal peoples. Broadly-speaking, only those fish which are commonly consumed or which are in other ways salient are accorded Nuaulu names. The traditional extraction of freshwater fish is reflected in a relatively more detailed lexicon for these fish given their total numbers than for marine forms. When faced with a marine fish for which there is no Nuaulu term, a Scpa or AM term may be used instead, if these are known. There are an indeterminate number of Sepa fish terms which are occasionally used, though they are not sufficiently shared to constitute what we might describe as 'assimilated' Nuaulu terms. For this reason they have not been listed in section 9.2, and some are presented separately in table 16. In this case it is difficult to know where the Nuaulu lexicon begins and that of other language groups end. If a term is introduced from another language it is clearly being used as a Nuaulu term. Whether a term is accepted as part of the Nuaulu corpus, must in the end depend on judgements as to general acceptability, shared use and conscious recognition that it is a Nuaulu term ('our word'), rather than one borrowed for the occasion. This matter is discussed in more general terms in chapter 2.6 of *The Cultural Relations of Classification*.

CHAPTER TEN

INSECTS

10.1 The insect fauna of south central Seram

The insect fauna of Seram is extraordinarily diverse and abundant. Forbes, 1885: 291, writing in his *A Naturalist's Wanderings*, reports that on Ambon alone insects - particularly beetles - are numerous and of great variety. By the end of the nineteenth century, Ribbe [Ribbe, 1892: 46] had recorded some 19,000 different species, including 10,000 butterflies, and reckoned to collect 300 insects daily. Today the number of certain species for all orders of insects must be many more than this. Clearly, only a very few specimens compared with the total number of known species were collected in the field, but even so they constitute the largest single group of specimens. Most importantly they include all common species encountered by the Nuaulu. A checklist of insect species for which specimens were collected in south Seram is presented in table 17.

10.2 Nuaulu categories applied to insects

Nuaulu terms for insects represent the largest single group in their animal inventory and provide the ethnographer with the greatest problems in presentation and analysis.

10.2.1 makasisi popole

The Nuaulu name for this predaceous insect, in AM 'capung', relates to its habit of feeding on insects (such as mosquitos) immediately above fresh water, and to the touching of the surface of the water or tips of plants with its tail. **Maka** comes from **makae** ('hard'), referring to the head; and **sisi**, meaning 'to tap, touch or scrape'. Thus **eresisi waene** is 'to scrape (the) water', where **ere** is a pronominal vowel prefix indicating a non-human actor. It was explained to me that the compound **makasisi** meant 'to dive' or 'to swoop', while **popo** meant 'to copulate with the water', referring to the way the insect hovers and touches the surface of the water with its cerci (extreme rear end).

The term is applied to fairly robust dragon-flies, commonly of the family Libellelidae, and to the thin-bodied damsel-flies frequently encountered in large numbers in the sago swamp forests of Somau (Tihun), at the Nua-Ruatan confluence. It may also be used more widely to include ant-lion flies such as *Myrmeleon*, which resemble dragon flies of the narrow bodied type.

Five types are recognised.

10.2.1.1 makasisi popolo msinae

Msinae = 'red', applying to a red-bodied dragon-fly.

10.2.1.2 makasisi popolo marae

Marae = 'blue-green (grue)', applying to greenish blue dragon-flies

10.2.1.3 makasisi popolo putie

Putie = 'white', applying to whitish or neutral-coloured dragon-flies

10.2.1.4 makasisi popolo masikune

Masikune = 'yellow', applying to yellow-winged dragon-flies. On one occasion was applied to the Protoneurid damselfly *Nososticta*.

10.2.1.5 makasisi popole pokotuene

Pokotuene = 'circular, round, spherical'; referring to the fact that the bodies of these damsel-flies tend to be round rather than long.

10.2.2 uri usue

Uri = 'banana, plantain' and **usue** = 'heart'; that part of the banana plant used as a vegetable. A term applied to small black earwigs *(DERMAP-TERA)* which feed on and infect bananas and plantains.

10.2.3 matamaine

Mata = 'eye'. A term applied to cockroaches (Blattidae) generally. Without further qualification it was used to describe specimens of both *Periplaneta australasiae* and *P. americana*. Two specific types of **matamaine** are recognised, referring to developmental stages.

10.2.3.1 matamai putie

Putie = 'white'

10.2.3.2 matamai reunosu

Reunosu = 'to shed'; referring to a white cockroach which has just shed its skin. Applied to both species of *Periplaneta* identified.

10.2.4 kinahorake (inae)

Kina may be from **kinate** (= 'glue, stuck, sticky'), referring to the difficulty of removing these insects from flesh once adhered. It appears to have no categorical implication. Applied to Panesthids (Blattidae).

TABLE 17 Checklist of insects recorded in the Nuaulu region of south central Seram, 1970-1975.

Species	Ecological zones					Nuaulu gloss
	1	2	3	4	5	
ODONATA - dragonflies and damselflies						
indet. spp.	-	-	-	+	-	**makasisi popole marae**
	-	-	-	+	-	**makasisi popole metene**
	-	-	-	+	-	**makasisi popole pokotuene**
ANISOPTERA - dragonflies						
Orthetrum villosovittatum	-	-	-	+	-	**makasisi popole msinae**
ZYGOPTERA - damselflies						
Protoneuridae						
Nosoticta sp.	-	-	-	+	-	**makasisi popole masikune**
DERMAPTERA - earwigs	-	+	-	-	-	**uri usue**
ORTHOPTERA						
Mantidae - praying mantids						
Hierodula sp.	+	+	-	-	-	**kau nimunone, kau kahe kamane**
Tenodera australasiae	-	+	+	-	-	**kau (mam) kahe kamane**
Blattidae - cockroaches						
Periplaneta australasiae	+	-	-	-	-	**matamaine**
?Periplaneta sp.	+	-	-	-	-	**matamai reunosu matamai putie**
Periplaneta americana						
Panesthidae						
indet. nymph	-	+	+	-	-	**kinahorake (inae)**
Phasmidae - stick and leaf insects	+	+	+	-	-	**kau ai otoe metene**
(Hierodula sp.)?	+	+	+	-	-	**kau ai otoe marae**
Xestophyrs sp.	-	+	+	-	-	**kauke**
indet. sp.	-	+	+	-	-	**kau ai otoe**
Platycrana viridana	-	+	+	-	-	**kau ai otoe**
Tettigonidae - bush crickets						
indet. sp.	+	+	+	-	-	**kau hatane**
Salomona marmorata ceramica	+	+	+	-	-	**kau nimunone**
Phyllophora bidentata	+	+	+	-	-	**kau kapine**
Hexacentrus sp.	-	+	+	-	-	**kau nesate**

Sexara coriacea	-	+	+	-	-	kau suto
indet. spp.	-	+	-	-	-	kau kasipi totue, kau putie
Gryllidae - crickets						
Cardiodactylus novaeguineae	-	+	-	-	-	kinapari
Gryllotalpidae - mole crickets						
Teleogryllus consimilis	-	+	+	-	-	kau tuaman anoe
Gryllacrididae						
Gryllaocris sp.	-	-	+	-	-	kau suto
Acrididae - locusts or short-horned grasshoppers						
Valanga nigricornis						kau nimunone
	-	+	+	-	-	kau nuhune (kau Matoke)
indet. spp.	+	+	+	-	-	kau hatu tinaie
	-	+	-	-	-	kau suenie
ISOPTERA - termites	+	+	+	-	-	tananae, tananae inae
Eutermes amboinensis						
Captotermes sp.						
Microcerotermes amboinensis						
ANOPLURIDA - sucking lice						utu
Haemotopinus suis apri - pig louse (*Sus scrofa*)	-	-	-	-	+	hahu utue (hahutue)
Pediculus humanus	-	-	-	-	+	utu
HEMIPTERA - true bugs						
Scuttelleridae						
Tectocoris diophthalmus	-	+	+	-	-	rikune, riku anae,
	-	+	+	-	-	riku pate, riku hanaie
Calliphora billardierii	-	+	+	-	-	kina puku paine
Cantao ?rudis	-	+	+	-	-	kapetite anae
Pentatomidae - stink bugs						
prob. *Oncomeris* (Tessanataminae)	-	+	+	-	-	rikune, rikune anae
Flatidae						
Euphanta monoleuca	-	+	+	-	-	rikune
Coreidae - squash bugs						
Mictis amboinensis	-	+	+	-	-	rikune
Mictis profana	-	+	+	-	-	riku wesie

Pyrrhocoridae - firebugs						
Dysdercus singulatus	-	+	+	-	-	makarota pina msahane
indet. sp.	-	+	+	-	-	kinapopote
Acanthiidae						
Cimex sp. - bed bug						
Tesseratomidae						
Plisthenes merianae	-	+	+	-	-	
Nogodinidae						
Papuana huebrieri	-	+	-	-	-	kitoe
Mindura sp.	-	+	-	-	-	kitoe
Physomenus grossipes	-	+	-	-	-	kitoe
Aspidomorpha areata	-	+	-	-	-	kitoe
HOMOPTERA						
Cicadidae						
Baeturia sp.	-	+	+	-	-	sisie, sisie putie
Platypleura sp.	-	+	+	-	-	hana-hana
(Dandubia) sp.	-	+	+	-	-	nai
indet.	+	+	+	-	-	sisie numa, sisie kauke
Fulgoridae						
Birdantis sp.	-	+	+	-	-	kau tuamane anoe, tananae inae
Gerridae						
immature sp. - pond skater	-	-	-	+	-	sa sahune
NEUROPTERA - lacewing flies						
Myremeleonidae						
Myrmeleon sp.	-	+	+	-	-	makasisi popole
COLEOPTERA - beetles						
Aspidomorpha creata	-	+	-	-	-	
Lymexylidae						
Atractocerus sp.	-	+	+	-	-	kunte inae
Elateridae - click beetles						
Lanelalus insularis	-	+	+	-	-	ka(u)petite
Agrynus resectus						
Buprestidae - metallic wood-borers						
Chrysodema sp.	-	+	+	-	-	riku pate
Chrysodema malacca						

Coccinellidae - ladybirds						
Coccinella repanda	+	+	+	-	-	**hite anae**
Anobiidae						
Anobium sp. - woodworm	+	+	+	-	-	**susine**
Lucanidae - stag beetles						
Lucanus sp.	-	+	+	-	-	**atori (nione)**
Scarabaeidae - scarab beetles						
Glycyphana perridis	-	+	+	-	-	**hite, kinapuku paine**
Oryctes rhinoceros - rhinoceros beetle	-	+	+	-	-	**atori (nione), sinne inae**
Cerambycidae - long-horned beetles						
Gnoma giraffa	-	+	+	-	-	**kamanahu metene**
Gnoma zonaria						
Glenea corona	-	+	+	-	-	**kamanahu putie**
Glenea sp.						
Mulciper linnaei	-	+	+	-	-	**kamanahu, sinne inae**
Hexamitodera semivelutina	-	+	+	-	-	**susine**
Coptocerus biguttatus	-	+	+	-	-	**susine**
Chrysomelidae - leaf beetles						
Altica sp.	-	+	+	-	-	**kinapopo(te) inae kinapopote (marae)**
Glyciphanax sp.						
Lampyridae - fireflies	+	+	+	-	-	**kinapopote**
Anthribidae						
Xenocerus semiluctuosus	+	+	-	-	-	**rikune ari ai kanapua**
Xenocerus sp.	+	+	-	-	-	**kinoi (metene)**
Curculionidae - weevils						
Rhynochophorus bilineatus	-	+	-	+	-	**kinapukune**
Rhynochophorus ferrugineus						
Calandra oryzae - rice weevil	+	-	-	-	-	**kinapukune anae**
Calandra granaria						
Passalidae						
Labienus moluccanus	+	+	+	-	-	**kinapari**
indet. (prob. *Podops* sp.)	-	+	+	-	-	**rikune, riku ai ukune**

HYMENOPTERA
Formicidae-Myrmicidae - ants

-	+	+	-	-	kinawerie
+	+	+	-	-	muna usite
-	+	+	-	-	kumte
+	+	+	-	-	sohone

Odontomachus simillimus	+	+	+	-	-	isanone
Solenopsis germinata						
Dolichoderus thoracicus	+	+	+	-	-	isanon metene
						isanon msinae
Monomorium spp.	+	+	+	-	-	uane utue

Vespidae
Ropalidia sp. - double-bodied wasp   +   +   +   -   -   imanine (inae)
Sphecidae - digger wasps
Sphex sp.   +   +   +   -   -   sene nan
Eumenidae
Eumenes - potter wasp   +   +   -   -   -   sene tuamane
Apidae - honey bees   -   +   +   -   -   inae kilalante
Apis indica
Apis dorsata
Apis florea
Apis cerana
LEPIDOPTERA - butterflies and
moths
Cossidae
Duomitus ceramicus   -   +   +   -   -
Morphidae
indet.   -   +   +   -   -   kori-kori marae
Pyralididae - flour and meal moths
Parotis sp.   +   +   +   -   -   marane tina totue

Papilionidae- swallow-tails and
apollos
Papilio fuscus   +   +   +   -   -   kori-kori metene,
           +   +   +   -   -   kori-kori nika
                            pante

Papilio ulysses
Papilio alcidinus
Papilio lorguinionas

	1	2	3	4	5	
Pachliopta polydorus	-	+	+	-	-	**kori-kori nika pante**
Arctiidae						
Maenas maculifascia	-	+	+	-	-	
indet. Lithosiinae						**kinopopote**
Noctuidae						
Lyssa docile	-	+	+	-	-	**kori-kori inahai kori-kori uri**
Danaidae						
Danaus juventa	+	+	+	-	-	**kori-kori metene**
Euploea						
Nymphalidae						
Inachis						
Lycaenidae	+	+	+	-	-	**kori-kori marae**
small *Lycaena*						
Dendorix ceramensis						
Arhopala ate						
Uraniidae						
Nyctalemon agathyrsus						
Pieridae						
Aporia crataegi						
Danaideu						
Priamus						
Helana						
Parthenos var *brunea*						
Alimena						
Erionata thrax	-	+	+	-	-	**kori-kori ikine**
Hypolimnas pandanus	-	+	+	-	-	**kori-kori metene**
Troides oblongmaculatus	-	+	+	-	-	**kori-kori (onate)**
Graphium codrus	-	+	+	-	-	**kori-kori onate**
indet.	+	+	-	-	-	**inahai**
Kallima pylarchus						
Sphingidae - hawkmoths						
Geometridae						
indet.	-	+	+	-	-	**kori-kori marae**
DIPTERA - two-winged flies						
Tipulidae - crane flies	+	+	+	-	-	**kunte inae**
Culicidae - mosquitos	-	-	-	+	+	**kunte**
Anopheles farauti moluccensis						
Anopheles (Myzomia) punctulatus						

	Zone 1	Zone 2	Zone 3	Zone 4	Zone 5	
Culex incl. *ceramensis*						
Tabanidae - horse flies						
Tabanus sp. (*fumipennis* group)	+	+	-	-	-	imanane
Tabanus sp. (nr. *furunculigenus*)						
Asilidae - robber flies						
Laphria sp.	+	+	-	-	-	sene nanan
indet. sp.	+	+	-	-	-	kinawerie inae
Syrphidae - hover flies						
Allograpta sp.						
Drosophilidae - small fruit flies						
Drosophila ananassae	+	+	+	-	-	mumne
						mum anae
Trypetidae	+	+	+	-	-	mumne
Micropezidae						
Muscidae						
Musca domestica	+	-	-	-	-	imanane
Orthellia timorensis	+	-	-	-	-	inapone (marae)
Calliphoridae - blow flies	+	+	+	-	-	atinotoe
Neriidae						
Nerius (?)nigrofuscus	+	+	+	-	-	uri usue, kumte inae
Nerius sp.						
indet. small flies	+	+	+	-	-	sohane inae
SIPHONAPTERA						
Pulex irritans - human flea	-	-	-	-	+	utu
indet. insect	+	+	+	-	-	monote inae
indet. small mites	+	+	-	-	-	nione inae

Key. Zone 1 = village; zone 2 = cultivated areas; zone 3 = forest; zone 4 = freshwater, including sago swamp; zone 5 = zooparasites.

10.2.5 The general category 'kauke'

Kauke is a generic term applied focally to mantids, stick-insects, crickets, bush-crickets, gryllids and grasshoppers; in other words, most orthopterans (though notably excluding cockroaches). The focal form, therefore, appears to be an insect with straight wings, a distinctive 'armoured' head, and with hind legs which are disproportionately large compared with body size. As with all terms applied to insects, where the number of encountered species vastly exceeds those which are terminologically distinguished, many species are simply labelled **kauke** without further nominal differentiation. Despite my own very considerable efforts to find individuals who could 'identify' certain specimens with proper names which they had learned, rather than *ad hoc* descriptions which they had invented, some could not be classified more specifically than **kauke** e.g. the leaf-mimic, *Xestophyrs.*

At least two categories labelled by the prefix **kau** are neither orthopterans, nor considered by the Nuaulu to be **kauke** in the broad sense: **kau atinotoe** (10.2.43), and optionally **(kau) kamanahune** (10.2.21).

10.2.5.1 kau kahe kamane, mam kahe kamane

The interiors of Nuaulu houses are still mostly lit using torches usually made of resin from the screwpine *Agathis dammara,* known locally as **kamane** [1]. As the resin burns it leaves an ash which unless periodically removed, dulls the flame and will eventually result in its extinction. The verbal phrase **kahe kamane,** 'to scrape the kamane', refers to this activity, which is usually performed with a small slither of sago palm leafstalk. The specific allusion in the name appears to be that the movement of the feet resembles the scraping of dammar.

Applied to mantids, including *Tenodera australasiae* and *Hierodula.*

10.2.5.2 kau ai otoe

Ai otoe is a short piece of wood or twig; thus a highly appropriate homolexeme for stick-insects. Applied to *Platycrana viridana,* and perhaps other species. Two types are recognised, but rarely distinguished.

10.2.5.2.1 kau ai otoe metene

Metene = 'black'

10.2.5.2.2 kau ai otoe marae

Marae = 'green-blue, grue'

Mantids and stick-insects are seen as being closely related forms, presumably on account of their both being camouflaged, long-bodied and long-legged insects. On several occasions I have heard people suggest that mantids are a type of **kau ai otoe,** but never that stick-insects were a type of **kau (mam) kahe kamane:**

kau ai otoe kau (mam) kahe kamane

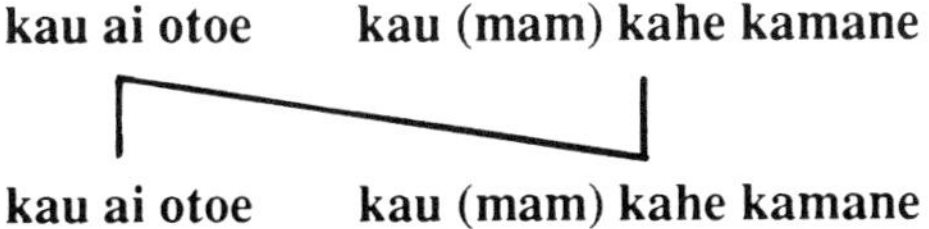

kau ai otoe kau (mam) kahe kamane

10.2.5.3 kau hatane

Hatane = 'the sago palm, *Metroxylon sagu'*. Applied to brown bush crickets of the family Tettigonidae, commonly found on dry banana leaves and on sago palms in cultivated areas.

10.2.5.4 kau (ni)munone

Nimunone (and possibly also **nemonone**)= *Melanthera biflora*, and other Compositae, known in AM as 'sunga-sunga'; a type of coastal weed of the upper shoreline of the coast of south Seram, on which this insect is found. Appears to be applied focally to the bush-cricket *Salomona marmorata ceramica* and grasshoppers such as *Valanga nigricornis*. One specimen of the mantid *Hierodula* (otherwise **kau (mam) kahe kamane**) was unaccountably given this label. Regarded as edible.

10.2.5.5 kau kapine

Kapine are decorated box-like containers made from sago leaf-stalk (**tope**) and pandanus leaf (**koae**). The larger boxes of this kind, generally associated with the Manusela area or with the islands off southeast Seram, are not made by the Nuaulu, although they are sometimes to be seen in Nuaulu villages. However, smaller containers in the same style, and used for betel-chewing requisites, are commonly found and manufactured. A term applied to the box grasshopper *Phyllophora bidentata*, and perhaps related species.

10.2.5.6 (kau) nesate, kau kasipi totue

Nesate is raw sago flour cooked in sago leaves, a common food on sago-extracting expeditions. **Kasipi** = 'manioc, *Manihot esculenta'*; **totue** = 'leaf'. Applied to large bush crickets, including *Hexacentrus*. The synonym is a reference to a common food of this cricket: manioc leaves. Regarded as **mine**, 'inedible'.

10.2.5.7 kau putie

Putie = 'white': bush crickets of indeterminate species, also nymphs of the family Tettigonidae (see 10.2.5.3).

10.2.5.8 kau suto

Suto = type of crustacean (chapter 12.2.4). Applied to the bush cricket *Sexara coriacea* and the gryllacrid *Gryllaocris*, and perhaps other gryllids; said to live on river banks and to be generally edible. Regarded as an omen (**monne**) by the clan Matoke: if a newly-born baby sees or is in close contact with a **kau suto** before its umbilicus has shrivelled-up, its chances of survival are thought to be slim.

10.2.5.9 kau tuaman anoe (kau tuamane)

Tuaman anoe, lit. 'inside (**anoe**) the earth (**tuamane**)', referring to the fact that this wingless mole-cricket lives in the ground and in low underbrush. Applied to *Teleogryllus consimilis,* and on one occasion to the lantern-fly *Birdantis.* Komisi says that it can only be caught at night.

10.2.5.10 kau nuhune, kau Matoke

Nuhune is a reference to certain birth rituals found in a number of clans, including Matoke. A brown grasshopper ritually restricted for the clan Matoke; hence the synonymous usage. Applied to the short-horned grasshopper *Valanga nigricornis.*

10.2.5.11 kau hatu tinaie

Tinaie = 'tree'; **hatu** = 'stone'. Indeterminate Orthoptera. Found in the ground surrounding places where villagers collect their water. At night they are attracted to village lamps.

10.2.5.12 kau suenie

Suenie = a type of bamboo (*Schizostachyum* spp.), on which these otherwise indeterminate Orthoptera are found.

10.2.6 tanana(e inae)

Inae = 'mother'; **tanana** = 'land, earth'; thus, 'mother of earth'. Applied to all species of termites, including *Captotermes,* the common *Eutermes amboinensis* known to nest in several kinds of tree including *Alstonia scholaris* and the sea hibiscus *Hibiscus tiliaceus*; and *Microtermes amboinensis,* known to nest in the coconut palm. The term was also applied to a specimen of the lantern fly, *Birdantis.*

10.2.7 (k)utu

Utu is the PAN root for the human head and body louse, *Pediculus humanus,* though it is sometimes extended to ticks and mites (table 22), and also to the human flea. Because of the irritation and inflammatory reaction (**heni-heni**) caused by lice, and the amount of time devoted to delousing and grooming, as well as the cultural importance attached to it, the life-cycle is well-understood. The eggs are termed **tenie.** These incubate after about eight days into the nymph (**utu putie, utu meni anae;** that is the 'white' or 'young' louse). Nymphal development, which takes a further 16-19 days results in an **utu metene** or **utu tia pokone** (lit. 'black louse' and 'full stomach', respectively). However, the transmission by lice of more serious diseases is not understood.

Although temporary relief from irritation can come from scratching or (as I have often seen) knocking the head with the back of a bushknife blade, lice are only effectively removed by careful and systematic grooming and then using a special comb (**senie**; made from bamboo with binding of *Lygodium scandens*). Cleaning is seldom complete and re-infestation rapid. Really effective treatment only comes with shaving the head, a practice which is now common for young children of both sexes after they have had their first hair-cutting ceremony. Lice are removed and crushed between the fingernails if they are large enough, or squashed on the head with the specially designed end of a comb. It is believed that squashed lice which remain in the hair can regrow. Young children will groom each other, or be groomed by their parents. As children grow older grooming between the sexes becomes an embarrassment, except between brothers and sisters. On marriage husband and wife will groom each other. Bonds expressed through grooming are clearly of great intimacy and psychologically significant, as indicated in the grooming of quite old unmarried adolescents by their mothers.

Lice, mites and ticks are familiar from other animals e.g. **ha(hu) utue** or **hahu te** [2] ('pig louse'), **asu utue** ('dog louse'), and so on.

10.2.8 rikune

Rikune is also the verb 'to gargle', but while homophonous the two terms are semantically distinct. As applied to insects, **rikune** refers to various kinds of true bugs and beetles. Without further qualification it was applied to the bugs *Oncomeris, Euphanta monoleuca, Mictis profana, M. amboinensis* and stink beetles

10.2.8.1 rikune ai kanapua

Lit. 'bug (which) eats **kanapua** bushes'. **Kanapua** = *Dryobalanops aromatica*, the camphor tree. Applied to the anthribid beetle *Xenocerus semiluctuosus*.

10.2.8.2 riku anae

Lit. 'child of the **rikune**'; applied to stink-bugs. Nuaulu believe that when these bugs urinate it makes your eyes hurt. More likely is the fact that they exude some noxious substance which when rubbed on the eyes makes them sting. Applied to the scuttellid bug *Tectocoris diophthalmus*.

10.2.8.3 rikupate

Pate is a tree of the primary rain forest. Applied, like **ansiha** (10.2.9), to the black scuttellid bug *Tectocoris diophthalmus*. It is said to have a nasty smell. Also applied to the metallic wood-borer *Chrysodema*, it is said to cause damage to house posts, and to eat rotting pork and venison.

10.2.8.4 riku hanaie

Lit. 'male bug'. Applied, as with **ansiha** and **rikupate**, to the scuttellid bug *Tectocoris diophthalmus*.

10.2.8.5 riku pina

Lit. 'female bug'; said to emit a fluid which stings the eyes and is generally unpleasant. It is unclear whether this is the female of **reku hanaie**, or whether it is a quite separate species.

10.2.8.6 riku wesie, riku ai ukune

Lit. 'forest bug'. Applied to the phytophagous squash bug *Mictis profana*. Regarded as a garden pest, found particularly on **karatupa** bushes (*Capsicum* spp.; chilli). The term **riku ai ukune,** lit. 'bug of the treetop', appears to be used synonymously.

10.2.8.7 riku nasate

Nasate = langsat fruit (*Lansium domesticum)* on which this bug is said to live and feed. Applied to *Plisthenes merianae*.

10.2.9 ansiha

Applied only to the scuttellid bug *Tectocoris diophthalmus*, commonly found on the branches of **kanapua** bushes (see 10.2.8.1) along the upper shoreline. **Ansiha** is regarded by most informants as a type of **rikune.**

10.2.10 makarota pina msahane

Pina msahane are 'female affines'. Applied to the cotton-stainer *Dysdercus singulatus,* and commonly found on the musk plant, *Abelmoschus moschatus.*

10.2.11 kitoe (inae)

Although **kitoe** also means 'a leak (in the roof of a house)', these appear to occur as semantically unrelated homophones. A brown beetle which eats and lives on the leaves of taro (*Colocasia esculenta)* when young; although also found on sweet potato (*Ipomoea batatas*): the taro beetle (technically a true bug) *Papuana huebrieri,* and also *Mindura* sp. Possibly also includes the true beetles *Aspidomorpha areata* and *Physomenus grossipes.* Edible.

10.2.12 sisie

Glossed in AM as 'uir-uir' or 'tunggaret', and not further reducible semantically. It is applied in its unqualified form to the cicada *Baeturia.* It is, however, used polysemously to refer to cicadas in general (distinguished by their distinctive sound-production) and to *Baeturia* in particular.

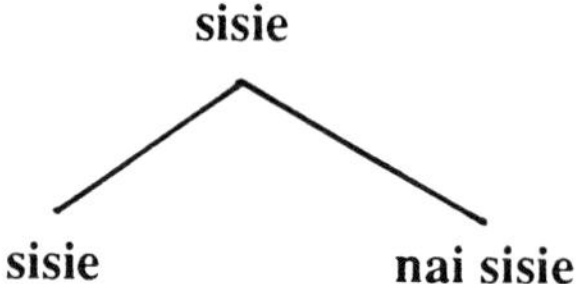

However, a number of specific types of **sisie** are recognised.

10.2.12.1 sisi putie

Lit. 'white cicada'; applied to *Baeturia.*

10.2.12.2 sisi kauke

On the meaning of **kauke** see 10.2.5. Not reported by R.B.

10.2.12.3 sisi numa

Lit. 'house cicada'. R.B. reports the possible synonym **sisi noine (mai numa)**, where **noine** = 'bud'.

10.2.12.4 sisi marae

Marae = 'green-blue (grue)'. Reported by R.B.

10.2.13 nai

The sound of this cicada is said to be 'the cry of the ancestors'. Possibly applied to *Dundubia.*

10.2.14 hana-hana

Applied to the cicada *Platypleura*, although no informant suggested that it might be related to either **sisie** or **nai**.

10.2.15 sa sahune

On the meaning of this term see chapter 9.2.39; applied to pond skaters of the family Gerridae.

10.2.16 hite

These beetles gather round lamps in the evening. The term was applied to the scarab beetle *Glycyphana perridis*.

10.2.16.1 hite anae

Lit. 'child of the **hite**'. The term is applied to the lady-bird, *Coccinella repaude*, which is thought eventually to grow into a **hite**.

10.2.17 kapetite

Kapetite = 'to beat, to tremble, to click'; the name is thus an appropriate one for these click beetles with their power of leaping when lying on their back. The term was applied to *Lanelalus insularis*. The term **kapetite anae** was on one occasion applied to the scuttellid bug *Cantao ?rudis*. It is interesting that occasionally such beetles should be described as **kau petite**, perhaps an understandable conflation arising from the application of a verbal form to insects where **kau** is a common prefix. Alternatively it may indicate some earlier perceived affiliation.

10.2.18 susine

The term is applied to 'woodworm' of the family Anobiidae. The Nuaulu distinguish clearly between the larvae (**susi ikine**: lit. 'small **susine**') and the mature beetle (**susi onate**: lit. 'large **susine**'). The latter is sacred (**monne**) for the clan Sonawe-ainakahata. Also applied to Cerambycid beetles which attack the clove tree: *Hexamitodera semivelutina* and the twig-borer *Coptocerus biguttatus*.

10.2.19 atori (nione), atoreo

The word **atori** is also used for 'scissors'. The comparison between the greatly developed mandibles of the male of this large stag beetle, and the results of their activity on leaves, and scissors is clear. The term for this well-known pest is generally qualified by **nione** (= 'coconut'; *Cocos* *nucifera*) indicating the coconut palm as the type habitat of this beetle, and the young palmate leaves as its principal food. It is a well-known pest. The term is glossed in AM as 'kwangwung', and was applied to the genus *Lucanus*. I have also heard the term applied to rhinoceros scarab beetles (see 10.2.20), *Oryctes rhinoceros*.

10.2.20 sinne inae

Sinne = 'house walls running from east to west'; thus lit. 'mother (**inae**) of the east-west house walls', but the term is also applied to a category of Nuaulu spirits. The term was applied to rhinoceros scarab beetles (*Oryctes rhinoceros*) which are believed to be the physical manifestation of spirits of the same name. The large grubs (**etine**) are much sought after as food. On one occasion the longhorn beetle *Mulciper linnaei* was described as **sinne inae.**

10.2.21 (kau) kamanahune

Kama(ne) is resin (see 10.2.5.1); **kamanahune** being the ghost or spirit of a person killed by falling from a tree in the throes of hunting cuscus. The prefix **kau** was only used on one occasion. The term refers collectively to long-horned beetles, which are regarded as edible. Two types of **kamanahune** are widely recognised, and perhaps up to four others less widely, though they are not generally distinguished terminologically. When they are, they are distinguished as follows:

10.2.21.1 kamanahu metene

Lit. 'black' long horn beetle. The term was applied to *Gnoma giraffa*.

10.2.21.2 kamanahu putie

Lit. 'white' long horn beetle. This was applied to *Glenea corona*.

10.2.22 kinapopote

For the possible meaning of **kina** see 10.2.4. **Popote** is a kind of jambu, an edible fruit of the genus *Eugenia*. The term, glossed in AM as 'kunung-kunung', is applied to all light-producing beetles and appears to include both light-producing click-beetles and true fire-flies (Lampyridae). However, on various occasions I have heard the term applied, in both a qualified and unqualified form, to the leaf-beetle *Altica* and to an Arctiid moth.

10.2.22.1 kinopopo marae, kinapopo inae

Marae = 'grue'; **inae** = 'mother of'. Both of the first two terms are applied to *Altica.* Glossed in AM as 'mai mai terong' (eggplant mite), on account of being a parasite of *Solanum melonogena.*

10.2.23 kinoi

Applied to fungus weevils of the family Anthribidae, some of which must be crop pests. At least three types are recognised.

10.2.23.1 kinoi metene

Metene = 'black'. Applied consistently to what were described as 'male' specimens of *Xenocerus.* 'Females' are said to be 'chocolate' in colour. *Xenocerus,* with its long antennae, resembles a long horn beetle, and it is some indication of the depth of Nuaulu entomological knowledge that they readily distinguish between the types terminologically.

10.2.23.2 kinoi msinae

Msinae = 'red'.

10.2.23.3 kinoi nikate

Nikate = 'pattern, drawing, design, decoration'.

10.2.24 kinapukune (kina puku hatane)

On the possible meaning of **kina** see 10.2.4; **pukune** = 'short'; **hatane** = the sagopalm, *Metroxylon sagu.* Applied focally and on most occasions to the sagopalm weevil, *Rhynochophorus bilineatus.* The eggs of this weevil are laid in the soft tissue at the base of the sago palm leafstalk, in lesions, in stored sago flour, in heaps of waste pith or in the rotting bole of felled stands. The larvae sometimes feed on the living stem and to this extent are a pest, but they are also much desired as food. Indeed, felled palms will often be deliberately left with some fibrous pith remaining to encourage the growth of larvae, which are later harvested (plate 15). Because the term **kinapukune** is so focally associated with *Rhynochophorus,* adjectival qualification is hardly necessary, although occasionally I have heard the term referred to as **kinapuku hatane,** in contrast to other forms of weevil for which the uninomial is sometimes employed.

The term **kinapukune** is applied by extension to many kinds of weevils, usually terminologically qualified by the name of the host on which they are parasitic, e.g. **kinapuku yakon,** 'maize weevil'. In some cases it is difficult to know whether such terms are accepted names shared by a large number of people, or spontaneous *ad hoc* descriptions. We can, however, distinguish the following.

10.2.24.1 kinapuku anae

Lit. 'child of the **kinapuku**'. The term is applied to the rice weevil, *Calandra oryzae*, known in AM as 'kipik beras'.

10.2.24.2 kinapuku wesie

Lit. 'forest **kinapuku**'. Many weevils found on forest trees. Appears to be partially synonymous with **riku wesie** (10.2.8.6), being used on those occasions were the focal reference group is **kinapukune** rather than **rikune.**

10.2.24.3 kinapuku paine

(**P)aine** = 'wound, scar'. The term is applied to the shield bug, *Calliphora billardierii,* and to the scarab beetle, *Glycyphana perridis.*

10.2.25 kinapari

A term applied to Passalid beetles living in decaying wood, for example *Labienus moluccanus.* It was also applied on one occasion to crickets of the species, *Cardiodactylus novaeguineae.*

10.2.26 kinawerie (kinawane)

Applied to large tree-living ants which are reputed to bite with great fierceness. In particular associated with symbiots of the ubiquitous ant-tree, **timnisie**, the Asdepiad *Hoya* with its distinctive large galls, which is common along the shoreline.

10.2.26.1 kinawerie metene

Lit. 'black **kinawerie**'.

10.2.26.2 kinawerie msinae

Lit. 'red **kinawerie**'.

10.2.26.3 kinawerie inae

Lit. 'mother of **kinawerie**'. The term was applied to robber-flies.

10.2.27 isanone

Refers specifically to a large-bodied but long-legged species of ant. Includes *Odontomachus simillimus, Monomorium* spp. and *Solenopsis germinata.* Two types are recognised, but the term may occasionally be extended to include **kinawerie**: hence **isanon kinawerie** (10.2.26).

10.2.27.1 isanon metene

Lit. 'black **isanone**.' Includes *Dolichoderus thoracicus.*

10.2.27.2 isanon msinae

Lit. 'red **isanone**'. Found on dry coconut leaves.

10.2.28 uane utue

Uane = 'rain'. Without further affix, the term **utu** is applied to lice. Refers to winged-ants of the genus *Monomorium,* known in AM as 'mai-mai hujan', 'rain bug'. These creatures are said to bite and are encountered in large numbers following heavy rain, when they swarm around kerosene lamps like kamikaze pilots. If they are deliberately killed it is believed that a headache will follow.

10.2.29 muna usite, mausite

Applied to a small red long-legged ant commonly found on domestic food scraps. Regarded as a type of **isanone.**

10.2.30 kumte

A large unidentified ant.

10.2.31 sohone

Unidentified ant.

10.2.32 imanine

Applied to wasps.

10.2.32.1 imanine (inae)

Lit. 'mother of **imanine**'. In both its qualified and unqualified forms this term is applied to double-bodied wasps, such as *Ropalidia,* which are well-known for their painful bites and the swellings which result.

10.2.32.2 imanine on

A term applied to a small red wasp.

10.2.32.3 imanine bunara

Bunara, the Nuaulu village of that name, gives its name to this large red wasp.

10.2.33 senete

Like **imanine,** this term is applied to wasps (and hornets), and there may be some overlap in content between the two categories, to such an extent that for some informants they are virtually synonymous.

10.2.33.1 sene ohu

Ohu refers to the dangerous bite of this large ground-living wasp which builds its nest in the earth.

10.2.33.2 sene nan

Nan(te) = 'sky', referring to the fact that this wasp 'lives in the above', building a large nest in roof-timbers of houses. The term is applied to the digger-wasp, *Sphex*.

10.2.33.3 sene tuamane

Tuamane = 'earth, ground (though in this sense) clay'. Applied to the potter wasp, *Eumenes*, the nests of which are a common sight in house roof spaces.

10.2.33.4 senete inae

Lit. 'mother of **senete**'. Possibly *Bombus*, or some other large bee.

10.2.34 (inae) (kilalante), kilalante inae, mui suane

Kilalante refers to 'honeycomb'; **lante** the split bamboo flooring of Nuaulu houses; hence, lit. 'mother of the honeycomb'. The terms **inae, kilalante, kilalante inae** and **inae kilalante** are apparently used interchangeably when in the forest, but are proscribed within the village where **mui suane** (**suane** = village ritual house, AM 'baileo') must be used. Applied to honey bees, a group which certainly includes *Apis indica*, but possibly *A. florea* and *A. dorsata* as well. In 1990 specimens of *Apis cerana* were obtained, a species hitherto unknown east of Sulawesi.

10.2.35 kori-korie

A generic term for butterflies and moths. The special problems relating to the application of adjectival qualifiers to **kori-korie** are dealt with in section 10.4.

10.2.35.1 kori-kori metene

Lit. 'black **kori-kori**'. The term was applied to the swallow-tail, *Papilo fuscus, Hypolimnas, Danaus juventa,* and other butterflies whose wing colour was predominantly black.

10.2.35.2 kori-kori masikune

Lit. 'yellow **kori-kori**'. Applied to some indeterminate butterflies of various species with yellow or ochrous wings, and some with yellow backgrounds and black markings; and to a small brown butterfly with eye-spots.

10.2.35.3 kori-kori waene anoe

Lit. 'in the river butterfly', referring to the fact that individual forms assigned to this category are found along rivers and streams. The term was applied to specimens which were described by other informants as **kori-kori masikune.**

10.2.35.4 kori-kori msinae

Lit. 'red **kori-kori**'. Applied to a butterfly whose wings were largely brown-black, but with some orange.

10.2.35.5 kori-kori putie

Lit. 'white **kori-kori**'. The term was applied to a small black and white butterfly. Possibly includes many Pieridae.

10.2.35.6 kori-kori nikate

Lit. 'patterned **kori-kori**'. The term was applied to a small black and white butterfly labelled by other informants **kori-kori putie** and to a butterfly with a black and white design on an orange and light brown ground.

10.2.35.7 kori-kori nusi

Lit. 'lime butterfly'; **nusi** = collective term for various species of the genus *Citrus*. A homophone is applied to a group which includes most of the larger birds of the sea coast plus some similar forms of the rivers and inland waters (4.2.2-8). In this context the term was applied to a large black and white butterfly.

10.2.35.8 kori-kori marae

Lit. 'green-blue butterfly'. The term was applied to a small blue butterfly and to a large butterfly with black and light blue wings and a light blue body with white fur. In so far as this term is applied to small blue butterflies (as is largely the case), it is focussed on Lycaenidae (e.g. *Dendorix ceramensis* and *Arhopala ate*). Possibly also includes some Morphidae and Geometridae.

10.2.35.9 kori-kori sinatane

Sinatane are nettles, including *Laportea stimulans* and *Procris frutescens,* upon which the caterpillars of this butterfly congregate. The term was applied to a small white form with a black rim around the wing edge and with a black dot in the centre of each wing. Komisi claimed that one specimen was also an immature **kori-kori marae.**

10.2.35.10 kori-kori tuamane anoe

Lit. 'in the ground butterfly', referring to the nesting place used by this butterfly. Tan wings with black markings, including eye-spots. Application uncertain.

10.2.35.11 kori-kori mani ahue

Lit. 'bird *mani* of the secondary forest butterfly'; **ahue** = secondary forest, referring to habitat. An account of the ecological characteristics of **ahue** is to be found elsewhere [Ellen, 1978c: 117]. The wings of this butterfly have a grey-brown ground with an orange rim.

10.2.35.12 kori-kori onate

Lit. 'big butterfly'. The term was applied to *Graphium codrus,* and a brown butterfly found on pathside leaves, but must undoubtedly be applied on an *ad hoc* basis to many large butterflies.

10.2.35.13 kori-kori uri (usue)

Lit. 'banana (or banana heart) butterfly', alluding to the principal food and habitat of this group of species. A term applied to the moth *Lyssa docile,* a brown-winged butterfly found on pathside leaves, a butterfly with yellow patches on a brown-buff ground, and a small brown butterfly of unknown colour but with eye-spots. Edible.

10.2.35.14 kori-kori ikine

Lit. 'small butterfly'; a term applied to *Erionata thrax,* though by extension no doubt to many small forms. C.f. **kori-kori onate** (10.2.35.12).

10.2.35.15 kori-kori nika pane putie

Nika(te) = 'pattern, design'; **pane** is 'pole' and also the Nuaulu name for *Anthocephalus macrophyllus,* a deciduous monopodial tree of the secondary forest. The term is applied to a butterfly with brown and white patterned wings. One informant described a specimen as a young **kori-kori marae.**

10.2.35.16 kori-kori nika pante

Nika as in **kori-kori nika pane putie** (10.2.35.15); **pante** is either 'left' or is a contraction of the term for the house verandah (**pantetane**). Applied to Papilionids such as *Papilio fuscus* and *Pachliopta polydorus*.

10.2.36 (kori-kori) inahai

This term is only occasionally prefixed by **kori-kori** and can generally be glossed by the English 'moth'. It refers to nocturnal Lepidoptera. Without further qualification it was applied to various species of moth.

10.2.36.1 inahai putie

Lit. 'white moth', a term applied to all such; perhaps focally Danaid milk-weed butterflies.

10.2.36.2 inahai metene

Lit. 'black moth'. This may be an *ad hoc* description rather than a recognised natural kind. If it is the latter it indicates a very open residual category.

10.2.37 (inahai) peni wanu

Peni is an animal category which includes pig, cassowary and deer [Ellen 1993: chapter 4.4]; **wanu** is 'sign', in the sense of augury. The term is applied to a moth which, if killed in the village at night, will ensure good hunting the following day. The use of the prefix **inahai** is variable.

10.2.38 (kori-kori) mara tina totue

Lit. 'ear leaf of cuscus'. The allusion is to the glossy wing sheen of this small white butterfly. The term was applied to *Parotis* and perhaps other related genera. It is regarded by all Nuaulu as a type of **kori-korie.**

10.2.39 nika pan (masikune)

Nika pante or **nikate** means 'pattern, design' or 'multi-coloured'; thus this is lit. 'the yellow pattern'. The term was applied in both its qualified and unqualified form to a moth with brown wings and yellow markings.

10.2.40 kunte

A term applied to all species of mosquito, and certain Micropezid flies, which to the untrained eye resemble mosquitos. Of the many species of mosquito found on Seram, *Anopheles (Myzomia) punctulatus* is a known carrier of malaria, breeding in stagnant pools between sagopalms. *Anopheles farauti* and *Culex ceramensis* were also collected in the field.

10.2.40.1 kunte inae

Lit. 'mother of the mosquito'. The term was applied to crane-flies, and the long-legged fly *Nerius(?) nigrofuscus*. These species bear a morphological resemblance to mosquitos despite being much larger, though application of the term to the Lymexylid beetle *Atractocerus* is less clear.

10.2.41 mumne, mum (n)anae(a)

These terms are applied to *Drosophila ananassae* and other fruit-flies, including the Trypetid flies found on jambus. They are identified by the Nuaulu through their small size and habit of hovering in large numbers (often in clouds) over perishable food and rotting flesh and vegetation, where they are known to lay eggs giving rise to grubs (**uneu**). They are said to be the responsible agents for infecting wounds. **Sohane inae** are regarded as being similar in this respect, but larger (10.2.47; also 14.1.1-4).

10.2.42 imanane

A term applied to the house-fly, *Musca domestica*, the horse-fly (including *Tabanus sp. near furunculigenus*), and sometimes to blow-flies.

10.2.43 inapone

A term applied to the Muscid *Orthellia timorensis*. This fly is also closely identified with egg-laying on food and the production of **uneu**.

10.2.44 kau atinotoe

Atinotoe refers to wood that has been cut. A term applied to blow-flies.

10.2.45 uri usue

Lit. 'heart of the banana'. See also 10.2.35.13. The term was applied to *Nerius(?) nigrofuscus*, often found on bananas (**uri**).

10.2.46 sohane inae

Lit. 'mother of **sohane**'. The term is applied to a small fly, whose larvae **sohane**, are found in abundance at old sago workings, over opened sago trunks (see 14.1.1-4). **Supana** is also used for insects described in this way, probably synonymously.

10.2.47 monote inae

Lit. 'mother of **monote**', where **monote** refers to a large category of plants which can be approximately, but inadequately, glossed as 'weed'. Large numbers of this indeterminate insect are to be found periodically in gardens.

10.2.48 nione inae

Lit. 'mother of the coconut'. This term, glossed in AM as 'mai-mai kelapa', was applied to small mites habitually found on neglected coconut flesh.

10.3 Caterpillars, grubs and maggots

In general terms, the Nuaulu may be said to understand the process of metamorphization and reproduction in insects, and the links between many imagos, their eggs and larvae (and in some cases, nymphs, pupas and cocoons) are known. The term **popo nukue nukue** is used to describe the process; **pokonukue** = 'parcel, bundle', 'to compress'. However, in particular cases detailed knowledge is lacking or contradictory, and other non-empirical explanations interposed. This is particularly interesting since Rumphius, writing towards the end of the seventeenth century, was inclined to accept a theory of spontaneous generation in insects which was even then being challenged by natural philosophers such as Redi. Despite his occasional protestations of agnosticism, he found it possible to record how cicadas, beetles and caterpillars arose spontaneously from bark or leaves of trees, or from dirt and decaying wood, and even provides 'experimental' evidence to support his position [Diakonoff, 1959: 128-9]. Clearly, Rumphius found further vindication of his views in the beliefs of the Ambonese, when he notes in a matter-of-fact way that they claim that cicadas arise out of (appropriately) 'caju lapy' (kayu lapi: 'cicada wood'), probably *Eonymous* [Wit, 1959: Lib. IV, 79].

The Nuaulu distinguish two kinds of larvae, approximately corresponding to the British English folk terms 'caterpillars' and 'grubs'. These are **une-une** and **etine** respectively. The number of terms listed here and habitually proferred by Nuaulu informants in no way reflects a complex knowledge of the relationship between imagos and mature forms. Additional terms may be generated on an *ad hoc* basis through addition of the suffix **anae** or prefixes **une** and **eti**, though most non-salient forms are assigned to an unqualified residual category of unnamed caterpillars or grubs.

10.3.1 une-une

These are generally the larvae of butterflies and moths, and although some of them may be eaten, it is only children who do so.

10.3.1.1 une nusi

The caterpillar of the butterfly **kori-kori nusi (totue)**, a term applied to specimens of the moth *Lyssa docile* (see 9.2.34.7). Regarded as inedible.

10.3.1.2 une asu

The term refers to the 'dogtail' (**asu** = dog; see 2.2.6) at the rear of this caterpillar, which may be either green or brown. It is inedible and most probably a hawk-moth grub (Sphingidae), which are distinguished by a horn or tubercle on the abdomen.

10.3.1.3 une putute

Putu = 'hot, very dangerous, blood of a person who has been murdered'. Appears to refer to at least two kinds of caterpillar. One is largely black with blue eye-spots, red spots and light green lines running laterally along the abdomen, and is blue at the base of each spine; may sting on contact with the skin, and is said to develop into **kori-kori metene**, a term applied to *Papilio fuscus, Hypolimnas pandanus* and *Danaus juventa* (10.2.35.1). The other is furry with an orange head and black and yellow banded abdomen, about 2 cms long and said to sting like **sina,** the nettle *Laportea decumana*.

10.3.1.4 une sinatane

Sinatane urone is the nettle *Procris frutescens*. The term for this black and white caterpillar, which is about 1 cm long, would therefore seem to allude to its stinging properties. Probably a synonym for 10.3.1.3.

10.3.2 etine

A term applied to the larvae of insects other than Lepidoptera, usually to those of beetles and bugs, which are generally regarded as edible.

10.3.2.1 eti ai

Lit. 'wood (or tree) larva'. This term seems to cover 'woodworm' in the English sense, that is *Anobium* (**susine:** 10.2.18), the larvae of metallic wood borers (**rikupate:** 10.2.8.3) and of *Duomitus ceramicus*, found in fallen trees and known in AM as 'olong-olong'.

10.3.2.2 eti sene, eti onate

Synonymous terms for the larvae of wasp imagos (**senete** = 'wasp': see 10.2.33. **onate** = 'large'). These are largish grubs found in dead tree trunks and are highly-valued as delicacies, alive or dead, cooked or raw. **Eti onate** may also be applied to the thick white grubs of large black beetles (possibly Dynastid beetles or large Passalids) found in old rotting tree-trunks.

10.3.2.3 eti hatane

Lit. 'sago grub'. this is the grub of the sago weevil, *Rhynochophorus bilineatus* (**kina puku hatane:** 10.2.23), and known in AM as 'ular sagu'. These are possibly those grubs most sought after as food (plate 15).

10.3.2.4 eti nione

Lit. 'coconut grub'. Small grub which infests coconut.

10.3.3 kusumun

The term for this larva may be derived from the AM word for *Phalanger*, 'kusu'. Applied to grubs of various crickets, cockroaches, weevils and beetles; all of which are said to be inedible.

10.3.4 uneu

Plural of **une-une**. A term generally applied to the larvae of flies, found on decomposing matter of all descriptions.

10.4 Some general features of Nuaulu knowledge and classification of insects

There is no single term for insects, or for any large group of invertebrates which have 'insect-like' or 'wug' characteristics, such as might be taken to include also forms such as spiders, scorpians, mites and centipedes. However, amongst these latter, mites and ticks are less obviously distinctive, often sharing Nuaulu categories with insects. It is for this reason that it has been convenient to include them here (10.2.7, 10.2.48).

Two large sub-divided categories stand out above all others in Nuaulu classification of insects. These are denoted by the terms **kauke** and **kori-kori**. **Kauke** is focussed on crickets, grasshoppers, mantids, bushcrickets, stick-insects and gryllids, but may include other forms. **Kori-kori** are distinguished by wing-form and are rigorously restricted to Lepidoptera. However, the category **inahai** (moths) is sometimes contrasted with **kori-kori**, sometimes treated as a sub-category. Nuaulu **inahai** is best regarded as

PLATE 15: Searching for the larvae of the sago weevil (**eti hatane**: *Rhyno-chophorus bilineatus*) in rotten sagopalm trunk, Rohua: 23 August 1973.

referring to nocturnal Lepidoptera, and thus excludes certain species (e.g. *Lyssa docile*) which would be treated as moths in English folk zoology.

There is considerable variability among informants on what names to apply to particular specimens, especially when it comes to the names for different colour phases or developmental stages. Given the enormous diversity of insect species, the Nuaulu must be highly selective with those that they individually name. If we examine a large category, such as **kori-kori, etine** or **kauke,** we may note that some species and genera are salient (well-known) and given specific names e.g. **kori-kori uri totue, eti hatane, kau tuaman anoe, kau kapine.** These types are known either because they are of economic significance, or because they are visually or behaviourally distinctive. Those ordinarily considered not to come within these quite specific categories are put in a residual category. It may be said that they are 'just **kori-kori'**, or they may be put into an *ad hoc* category which locates them but does not make them into a true natural kind (e.g. **kori-kori wesie** : 'butterflies of the forest'). Finally, similarities in habits and form may make it possible for some unfamiliar types to be placed in existing categories, by their legitimate extension.

A second kind of category applied to insects are covert or pseudo-covert groupings. Thus, the cicadas **sisie, nai** and **hana-hana** are commonly grouped together on account of their sound production and morphology. They, and other groups are pseudo-covert in the sense that on any one occasion **hana-hana** might be said to be a type of **sisie,** or **sisie** and **hana-hana** might be said to be types of **nai.** We can distinguish five main pseudo-covert groups of this kind:

1. **sisie - nai - hana-hana**: cicadas
2. **kinawerie - isanone - uma usite - kumte**: ants
3. **senete - inae kilalante**: wasps and bees
4. **imanona - mumne - inapone - uri usue - atinotoe**: true flies
5. **etine - une-une - kusumun**: caterpillars, grubs and maggots

Some pseudo-covert groups are more *ad hoc* than this. For example, on one August evening during 1973 Naupate brought some specimens of the sago weevil *Rhynochophorus bilineatus* to my hut. In answer to my question as to whether there were other kinds of **kinapukune** he listed the following and indicated that there were also others:

kinapuku wesie
kinapuku hatane
kinapari
rikupate
rikune

The remaining categories for insects are either not further sub-divided at all (e.g. **susune**), or are only sub-divided into a few terminal categories (e.g. **rikune, matamaine**).

10.5 The social and economic significance of insects

Some of the larger crickets, grasshoppers and locusts (**kinapari, kinoi, kinapukune, rikune**) are frequently roasted and eaten, but more often by children in the context of play. Grubs of certain species (e.g. *Rhychophorus bilineatus*) are highly-valued and sought-after. Children may sometimes also eat certain types of caterpillar, but generally the rule is **une-une** and **kusumun** (inedible) : **etine** (edible). Grubs may be eaten live or dead, uncooked, boiled or roasted. Apart from this they are not prepared further, for example by skinning and disposal of intestine, as mentioned by Wit, 1959: Lib. V, p. 201 for the Ambonese. The range of edible insects is ritually restricted only in a very limited way (e.g. **kau Matoke**), but many more are not eaten simply because they are too small, or because the tradition does not exist. It cannot, however, be claimed that insects represent a significant part of Nuaulu diet, even if in the absence of other forms of animal protein or under famine conditions they may occasionally prove critical.

Honeycomb is collected from at least one species of bee, and the propolis is used as an adhesive, for example in the manufacture of the **orane** ritual headress. Although the Nuaulu consume some honey themselves, current high interest in honey-collecting reflects much more market demand outside the immediate area. Some insects are important ingredients in hunting magic, eaten in order to make the 'liver hot', to make men more efficient hunters. These potions are generally clan-specific, and administered with an appropriate verbal formula. For example, in the clan Somori, two types of large ant (**kinawerie msinae**), wasps (**senete**) and two varieties of double-bodied wasp (**imanine**) are used in this way. The last of these is administered mixed with *Capsicums*. It may be significant that these are all predatory social insects, which can inflict a nasty bite. The imagery of the *Capsicums* will be obvious. Among the hunting potions of the clan Nepane-tomoien is one containing the Muscid fly, **inapone** (10.2.43).

Insects are recognised as pests, although little is done to control them. The common white wood-worm (**susine**) is among the most common and visible, and is widely-known to be responsible for killing off clove-trees. However, the Nuaulu are new to clove-cultivation and do not appear to have suffered through epidemics of clove tree parasites. More familiar and better understood are bugs such as **kitoe**, parasitic on edible aroids and sweet potato.

Insects feature prominently in children's play activity. Not only are they cooked and eaten in imitation of adult food preparation, but may be the source of endless hours of amusement. Large butterflies, for example, are attached by thread to the thorax and kept as short-term pets, almost like animate kites.

Notes to Chapter 10

1   *AM = 'dammar'. A. dammara = A. alba* in some earlier publications.

2   Also known as **hahu wanane. Numa wanane** is a temporary shelter built as a resting place along a forest path, so **wanane** may here indicate the pig as the resting place of the louse.

CHAPTER ELEVEN

MOLLUSCS

11.1 The mollusc fauna of south central Seram

The mollusc fauna of the central Moluccas (particularly marine gastropods) is rich, and has been well-documented ever since Rumphius first described the main forms. At the present time many hundreds of certain gastropod species are known, including in excess of 104 terrestrial species. A few cephalopods and polyplacophora are known, named and used by the Nuaulu but, apart from *Nautilus*, are not seen as being related to molluscs. Shelled gastropods, bivalves and *Nautilus* are grouped together by the Nuaulu in a named and well-defined category: **nunu.** Squid and octopus are related only through their similarity with *Nautilus*. Of the known species which might be theoretically grouped by the Nuaulu under the label **nunu,** only 39 were observed and collected during fieldwork in the Nuaulu area. The difference between the number of species collected and those known zoologically from the area may be accounted for by relative geographical distribution. The south coast of Seram between Elpaputih Bay and Teluti Bay is relatively uniform and exposed, and without the reefs which provide a wide range of different niches. On the other hand, the restricted use of the sea by the Nuaulu may also be reflected in the low index.

A checklist of molluscs collected in the Nuaulu area is presented in table 18. Species identifications compared with Nuaulu designations applied to actual specimens collected are set out in table 19.

11.2 Nuaulu categories applied to gastropods and bivalves

11.2.1 nunu ai otoi.

Ai otoi is a short piece of wood or twig and in this context it appears to be an allusion to the shell length of this marine gastropod, on the basis of a superficial resemblance. Applied to all species of the genus *Conus,* and approximately equivalent to AM use of 'congkak berjari'.

11.2.2 nunu inanaie, nunu inane aie

Ina = 'mother', **aie** = 'base, ground'. The term, which glosses with AM 'lola' is applied to all species of the genus *Trochus*.

In many coastal areas of the Moluccas *Trochus* shell was traditionally used for the manufacture of bracelets and pendants. However, there is no existing Nuaulu tradition and it is unlikely that there ever has been in the historical period. Shell ornaments collected between 1970 and 1971 were said to have come from Teor and other islands off the southeast of Seram.

11.2.3 nunu iwa

Etymology unknown. The term, which glosses the AM term 'keong lola', is applied to *Trochus niloticus* and *T. flammulatus*, thus overlapping the content of **nunu inanaie**. The overlap appears to result from the tendency to use the term **nunu inanaie** to refer to smaller specimens and **nunu iwa** to refer to larger ones with red markings. If this is so, then the terms do not refer to true 'natural kinds', as Bulmer would use this term. The usage also makes sense of informants statements that the shell of **nunu iwa**, but not **nunu inanaie,** is occasionally sold in Sepa. The flesh is eaten.

11.2.4 nunu hihikuro

Etymology unknown. The term, which glosses with AM 'mata bulan' ('eye of the moon'), is applied to *Turbo setosus* and *T. argyrostomus*. One informant applied the term to *Drupa morum* (see 11.2.8).

11.2.5 nunu mata putie

Mata putie = 'white eye'; an allusion to the markings of this shell. Applied to *Turbo porophyrites*.

11.2.6 nunu mata ipole hanaie

Mata = 'eye'; **ipole** means 'to beat out (as in washing on a flat stone)'; **hanaie** = 'male'. Applied to *Lunella coronatus*.

11.2.7 nunu mata ipole pina

Pina = 'female'. Both **nunu mata ipole hanaie** and **nunu mata ipole pina** superficially resemble each other, except the latter is much smaller. The term **nunu mata ipole pina** is applied to *Angaria lacinatus*.

TABLE 18 Checklist of molluscs recorded in the Nuaulu region of south central Seram, 1970-1975.

Species	Ecological zones			Nuaulu gloss
	1	2	3	
AMPHINEURA				
- chiton	+	-	-	
GASTROPODA				
PROSOBRANCHIA - mostly				
marine spiral shells				
Conus g. lividus	+	-	-	**nunu ai otoi**
Conus cf. pertusus	+	-	-	**nunu ai otoi**
Conus cf. monachus	+	-	-	**nunu ai otoi**
Trochus tenebrica	+	-	-	**nunu inanaie**
Trochus niloticus	+	-	-	**nunu iwa**
Trochus flammulatus	+	-	-	**nunu iwa, nunu inanaie**
Trochus costatus	+	-	-	**nunu inanaie**
Turbo setosus	+	-	-	**nunu hihikuro**
Turbo argyrostomus	+	-	-	**nunu hihikuro**
Lunella coronatus	+	-	-	**nunu mata ipole hanaie**
Turbo porophyrites	+	-	-	**nunu mata putie**
Angaria lacinatus	+	-	-	**nunu mata ipole pina**
Drupa morum	+	-	-	**nunu unie, nunu hihikuro, nunu tapako**
Drupa spathulifera	+	-	-	**nunu tapako**
Murex tribulus	+	-	-	**nunu unie**
Septaria sanguisaga	-	+	-	**nunu wenate**
Septaria sp.	+	-	-	**nunu hua inate**
Cypraea arabica	+	-	-	**nunu hun**
Cypraea carneola	+	-	-	**nunu hun**
Cypraea lynx	+	-	-	**nunu hun**

Clypeomorus subbrevicula	+	-	-	**nunu oiro**
Clithon angulosa	+	-	-	**nunu tari**
Clithon(?) subsulcata	+	-	-	**nunu marane**
Latirus turritus	+	-	-	**nunu sesu nuae**
Barbatia fusca	+	-	-	**nunu hua inate, nunu katenane, nunu wae inate**
Thiara amarula	+	-	-	**nunu saun kanie**
Pisania sp. cf. crenilabrum	+	-	-	**nunu asu meie**

PULMONATA - land and
      freshwater snails

Chloritis mima	-	+	-	**nunu kinihane**
Chloritis ungulina	-	+	-	**nunu kinihane**
Neritodryas cornea	-	+	-	**nunu marane**
Nerita antiquata)	-	+	-	**nunu marane**
Achatina fulica	-	-	+	**nunu keon**
Amphidromus(?)	-	-	+	**nunu weri-weri**
Nanina citrina	-	-	+	**nunu kinihane**
Nania aulica	-	-	+	**nunu kinihane**
Melanoides (?)granifera	-	+	-	**nunu saun kanie**
Melanoides (?)punctata	-	+	-	**nunu seu waene**
Melania hastula	-	+	-	**nunu sesu waene**
Paludina	-	+	-	**nunu sesu waene**
Vivipara javanica	-	+	-	**nunu sesu waene**

PELECYPODA -mainly
bivalves

Remis sp.	+	-	-	**nunu mara nanate**
Remis (Corbicula) sp.	+	-	-	**nunu kakante**
Tridachnes elongata	+	-	-	**nunu kakante**
Tridacna maxima	+	-	-	**nunu moi ika**
Periglypta reticulata	+	-	-	**nunu moi ika**
Ostrea spp.	+	-	-	**mutiara inae**
Saccostrea cucullata	+	-	-	

	Zone 1	Zone 2	Zone 3	
Pinctada maxima	+	-	-	
	+	-	-	**nunu sikewe aie**
CEPHALOPODA				
Nautilus pompilius				
and prob. other species (*tenuis, major*)	+	-	-	**nakatua saha**
-squid				
Loligo edulis	+	-	-	**sonto, sonto hatu**
-cuttlefish				
Sepia pharaonis	+	-	-	
-octopus				
Octopus aegina	+	-	-	**urita**
Cistopus indicus	+	-	-	

Key. Zone 1 = marine; zone 2 = freshwater; zone 3 = terrestrial.

11.2.8 nunu unie

Unie, which may be glossed as 'bone', refers to the spikes on the shells labelled with this term. Glossed in AM as 'bia berduri', the term is applied to specimens of *Drupa morum* and *Murex tribulus.* The focal genus is almost certainly *Murex.*

11.2.9 nunu tapako

Tapako = 'tobacco', *Nicotiana tobacum*; but why this gastropod should be so named is unclear. It may be a corruption of **tanapaku,** 'the antlers of a young deer'. If this is so then it closely resembles the AM gloss 'cangkok tanduk' and reinforces the possibility of terminological confusion with **nunu unie.** The term is applied to *Drupa spathulifera* and *D. morum.* Both species are used in the manufacture of betel lime.

11.2.10 nunu hua inate, nunu (wae) inate

These are apparently synonyms. The meaning of **inate** is 'cut open half', and **hua** is a classifier for fruit, giving us something like 'the cut open half of fruit'; **wae** refers to freshwater (**waene**). Applied to *Septaria sp.* and *Barbatia fusca.*

11.2.11 nunu hun(i), nunu Tihun

Synonyms, the first apparently a contraction of the second. Tihun is the name give to the extensive sago swamp forest at the mouth of the river Ruatan by the people of Sepa, and known more usually to the Nuaulu as Somau. The terms, glossed in AM as 'bia kepala kambing', are applied to the

TABLE 19 Species identifications compared with Nuaulu terms applied to 66 gastropod specimens.

Species	N	1	2	3	4	5	6	7	8	9	10	11	12	13	14	15	16	17	18	19	20	21
Conus lividus	1	1																				
Conus pertusus	4	4																				
Conus monachus	2	2																				
Trochus tenebrica	4		4																			
Trochus niloticus	4			4																		
Trochus flammulatus	2		1	1																		
Trochus costatus	1		1																			
Turbo setosus	4				4																	
Turbo argyrostomus	1				1																	
Lunella coronatus	1						1															
Turbo porophyrites	1					1																
Angaria lacinatus	1						1															
Drupa morum	3				1			1	1													
Drupa spathulifera	1								1													
Murex tribulus	1							1														
Septaria sanguisaga	1																					1
Septaria sp.	2									1	1											
Cypraea arabica	4											4										
Cypraea carneola	2											2										
Cypraea lynx	1											1										
Clithon angulosa	2												2									
Clithon subsulcata	2													2								
Latirus turritus	1																1					
Barbatia fusca	3									1	1							1				
Thiara amarula	2														2							
Pisania crenilabrum	2																		2			
Chloritis mima	1																			1		
Chloritis ungulina	1																			1		
Nanina citrina	2																			2		
Neritodryas cornea	1												1									
Nerita antiquata	1												1									
Amphidromus	1																				1	
Melanoides granifera	2													2								
Melanoides punctata	2															2						
Melania hastula	2															1						1

TOTAL   66

Key N = total number of specimens identified by informants. 1. **nunu ai otoi** 2. **nunu inanaie** 3. **nunu iwa** 4. **nunu hihikuro** 5. **nunu mata putie** 6. **nunu mata ipole** 7. **nunu unie** 8. **nunu tapako** 9. **nunu hua inate** 10. **nunu wae inate** 11. **nunu hun** 12. **nunu tari** 13. **nunu marane** 14. **nunu saun kanie** 15. **nunu sesu waene** 16. **nunu sesu nuae** 17. **nunu katenane** 18. **nunu asu meie** 19. **nunu kinihane** 20. **nunu weri-weri** 21. **nunu wenate** cowries *Cypraea arabica, C. carneola* and *C. lynx.*

11.2.12 nunu oiro

Etymology unknown; applied to a gastropod of the inter-tidal zone: *Clypeomorus subbrevicula.*

11.2.13 nunu tari

Etymology unknown; applied to *Clithon angulosa.*

11.2.14 nunu marane

Marane = marsupial cus-cus or *Phalanger,* with which this natural kind is compared. The allusion is to both the coloration of this freshwater snail and its habit of climbing trees and grass stems at the waters edge. The AM gloss is 'siput rakus'; applied to *Clithon (?)subsulcata, Neritodryas cornea* and *Nerita antiquata.* The flesh is edible and the shell used in the manufacture of betel lime.

11.2.15 nunu saun kanie

Kanie ='seed (of the) **saun(e)**, (the screw pine *Pandanus conoideus*). Applied to the freshwater snails *Melanoides (?)granifera* and *Thiara amarula.*

11.2.16 nunu sesu waene

Sesu derives from **sesunue** meaning 'pin' or 'needle'; an obvious allusion to the shape of this freshwater (**waene**) gastropod. The term is applied to *Melanoides (?)punctata* and *Melania hastula.* When boiled these snails are considered edible.

11.2.17 nunu sesu nuae

See 11.2.16, except that in this case the reference is to a marine (**nuae**) gastropod; applied to *Latirus turritus.*

11.2.18 nunu katenane

Etymology uncertain, although **katonane** is the resonant hollow clucking sound made by flicking the tongue against the roof of the mouth. Applied to the bivalve *Barbatia fusca.*

11.2.19 nunu asu meie

Asu = 'dog'; **meie** may possibly be derived from **mei-** = 'tongue'; giving us 'dog tongue shell'. The term is applied to *Pisania sp. cf. crenilabrum.*

11.2.20 nunu kawasa (kuasa)

Etymology unknown, except that the same word occurs as a male personal name with mythical connotations, when it is regarded as sacred [Ellen, 1983]. The homophone also means 'power' (as in the AM gloss, from which it may be derived), though whether the words are homonymous is unclear. This is an edible freshwater snail, probably *Thiara amarula.*

11.2.21 nunu wenate

The term, which may be a contraction or mis-hearing of **wae onate** ('big river'), is applied to the freshwater snail *Septaria sanguisaga.* The flesh is edible and the shell used in the manufacture of betel lime.

11.2.22 nunu pukune

Pukune = 'short'; applied (apparently synonymously) to *Clithon (?)subsulcata.* See 11.2.14.

11.2.23 kinihane

This term (the meaning of which is not further reducible) is applied to all land slugs.

11.2.24 (nunu) kinihane

Lit. '(shell) slug'; applied to the land snails *Chloritis ungulina, C. mima, Nanina citrina* and *N. aulica.* Some confusion may arise from the common reduction to the uninomial **kinihane,** which as we have seen (11.2.23) also means 'slug', and the use of **kinihane** to mean 'snail shell', as in **kuma kinihane,** that is hermit crabs inhabiting a snail shell (chapter 12.2.3). Snails in this category are not eaten as they are considered poisonous; neither is the shell used in the manufacture of betel lime, perhaps as an extension of this practice, or more rationally because the shells are thin and rarely available in sufficient quantities to make their collection worthwhile. Shells of *Chloritis ungulina* are found in large quantities in the surface litter of limestone caves in the vicinity of Rohua also used when hunting bats.

11.2.25 (nunu) matakopui

Matakopui = 'decaying (or septic) eye' or 'sleepy dust', a reference to the sticky emission of this land snail found on trees and leaves.

11.2.26 (nunu) weri-weri

This term (which may be an AM introduction, or from some other coastal language) seems to be a collective term applied to all land snails, including **kinihane** and **matakopui.** The term was applied to a single specimen of *Amphidromus*; found among trees and grass along the beach.

11.2.27 nunu keon

From AM 'keong' for snail. Applied to *Achatina fulica*. This large snail has been introduced from east Africa through human agency, and is widespread throughout the tropics. In the early seventies I came across no specimens at all in the Nuaulu area; by 1990 it had become extremely common in secondary forest and on garden land where it was beginning to cause significant damage. According to Fred Naggs (pers. comm.), a population explosion following initial introduction is typical and is usually followed by a population crash, apparently due to a parasite burden. The term **nunu nisi** (reported by R.B.) may be applied to this species.

11.2.28 nunu kakante

Kakante may be the noun form of the verb **kaka,** 'to lift, to pick up'. The term is applied to the bivalve *Tridachnes elongata,* and perhaps also to *Remis (Corbicula).*

11.2.29 Nunu mara nanante

Mara nanante is a corruption of **mara makinete** (the male *Phalanger maculatus*). However, the term is so similar to **kakante,** and the apparent application to *Remis* so close, that they may refer to the same category.

11.2.30 nunu mua ika, nunu moi ika, mai ika (nunu uma ika, nunu umai ika; R.B.)

Mua and **mai** may be corruptions of **moi** ('respect'), **uma** or **umai** the same of **numa** ('house'); **ika** is 'fish'. We thus have something like 'fish to be respected' or (more probably) 'fish house'. The term, glossed in AM as 'keong kepala kambing' is applied to the clams *Tridacna maxima* and *Periglypta reticulata*. The flesh is eaten, and the shell used in the manufacture of betel lime, although it is not universally considered suitable as its thickness often prevents its reduction to lime given the primitive burning methods employed by the Nuaulu. Large clam shells are used as containers, for example to hold water used when sharpening bushknives. They are a common sight placed outside the door of a Nuaulu house next to the whetstone. Smaller shells are used as domestic scrapers and spoons: to clean containers, scrape fireplaces and to ladle sago porridge.

11.2.31 nunu sikewe aie

Sikewe = 'taro', *Colocasia esculenta*; **ai-** = 'foot, leg': giving us 'taro foot shell'. The identification of this marine shellfish is not known.

11.2.32 mutiara inae, supu putie (arch.)

Mutiara is identical with AM, meaning 'pearl'. So we have, quite literally, 'mother of pearl', referring to oysters of the genera *Ostrea, Saccostrea cucullata* and the pearl oyster *Pinctada maxima*. Not present in Nuaulu area; for the most part known by reputation only, the Aru islands of the southeast Moluccas being an historically important pearl fishery.

11.2.33 nakatua saha

Nakatua ('cockatoo': chapter 4.2.26) is an allusion to the beak-like shell of this cephalopod. **(M)saha** is the referential prefix for males who are married but not yet with offspring. The term refers to the genus *Nautilus* (most usually *N. pompilius*), the shell of which is used to decorate the ritual head-dress (**orane**) and shields (**aniaue**) [see plate 16 here, and Ellen 1993: frontispiece and plate 1.6e].

11.2.34 nunu purai

This is the only non-mollusc to be included in the category **nunu,** and refers to the barnacle.

11.2.35 sonto

Cognate with AM 'sontong'; applied to all types of squid and cuttlefish known to the Nuaulu. Squid, like marine fish, was until recently rarely caught by the Nuaulu themselves, and the specialised techniques required are still known and used by only a handful of individuals. Squid is occasionally purchased from non-Nuaulu. In 1990 a distinction was being drawn by some individuals between between **sonto** (optionally **sonto (a)abiasa**) and **sonto hatu** (optionally **sonto batu.** The latter is a larger species, found for example off the coast of Saparua. Note the interchangeability of AM and Nuaulu words, which presumably reflects the recency of the innovation.

PLATE 16: Shell (above), section (centre) and animal (below) of *Nautilus pompilius*. These wash drawings are the presumed models for the engravings in Rumphius's *Amboinsche Rariteitkamer*. Reproduced from Van Benthem Jutting [1959: photo 17]; original in Royal Library, The Hague.

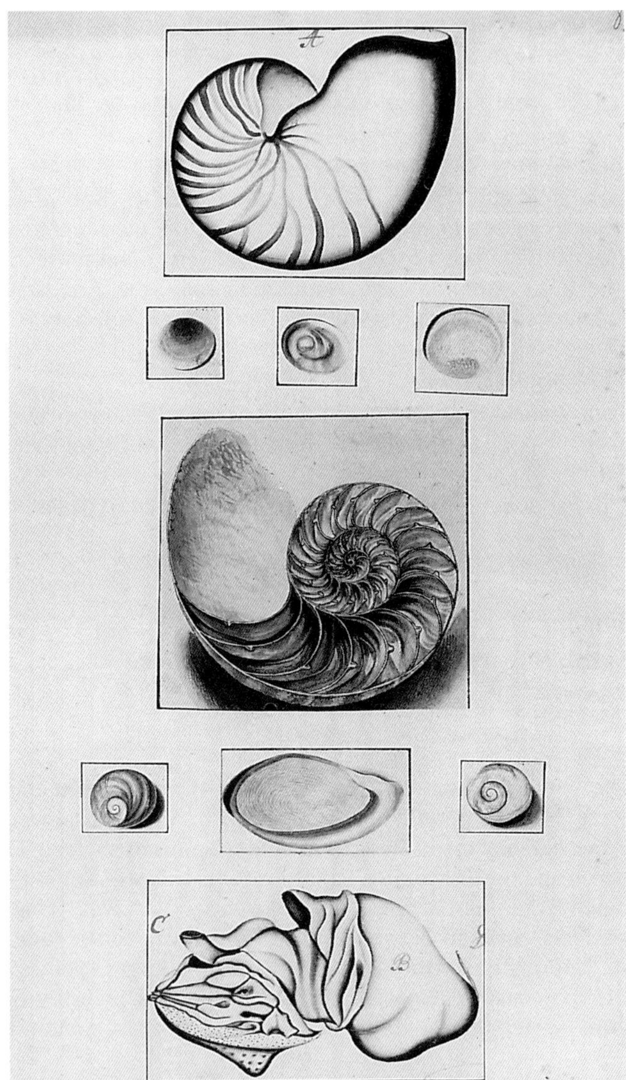

11.2.36 urita

Cognate with AM 'gorita'; applied to octopus, which is only rarely seen along this part of the coast of south Seram.

11.3 Non-basic categories for molluscs

All molluscs known to the Nuaulu, with the exception of squid and octopus, are placed in the well-defined category **nunu,** 'animals with shells'. Barnacles (**nunu purai**) are also included in the category on the grounds of possessing a hard outer casing and generally similar lifestyle. *Nautilus* is quite firmly assigned to the category **nunu** by virtue of its characteristic shell. While neither squid nor octopus are regarded in any sense as **nunu,** their similarity with *Nautilus* (tentacles, beak, eyes, ink) leads them to be loosely associated with the category on morphological grounds; they are 'linked' in Hunn's sense. Squid has a similar loose association with **ikae** on the basis of habitat and the similar methods employed to catch them. Land snails are only rarely identified terminologically as **nunu,** but they are generally assigned to the category. Their association with slugs (**kanihane**) may partly explain this marginal position. The two main types of land snail recognised by the Nuaulu (**matakopui** and **nunu kinihane**) are grouped together into a semi-covert category, sometimes labelled **weri-weri.** To this we should now perhaps also add the relative newcomer **nunu keon.**

Among 'core' **nunu,** we can infer the existence of two intermediate categories based on close physical resemblance: **nunu sesu** (11.2.16-17) and **nunu mata ipole** (11.2.6-7). More broadly, Nuaulu sometimes make an intermediate distinction on the basis of habitat, between **nunu waene** ('freshwater shellfish') and **nunu nuae** ('marine shellfish'). This contrast does not, of course, fully partition the category, excluding land snails. Together with the fact that these latter are never - for example - described as **nunu tanane,** this further underlines their marginality.

The relationships between the more inclusive non-basic categories discussed in this section are illustrated in figure 14.

11.4 The social and economic uses of molluscs

Various social and economic uses of molluscs and their allies for the Nuaulu have been dealt with in the individual entries for terminal categories; these are more systematically and comprehensively set out in table 20. Conch shells, *Choronia tritonis,* used as horns in west Seram, are known to the Nuaulu, but not used by them. The use of shells in the manufacture of

FIGURE 14 Relationships between the more inclusive Nuaulu categories applied to molluscs.

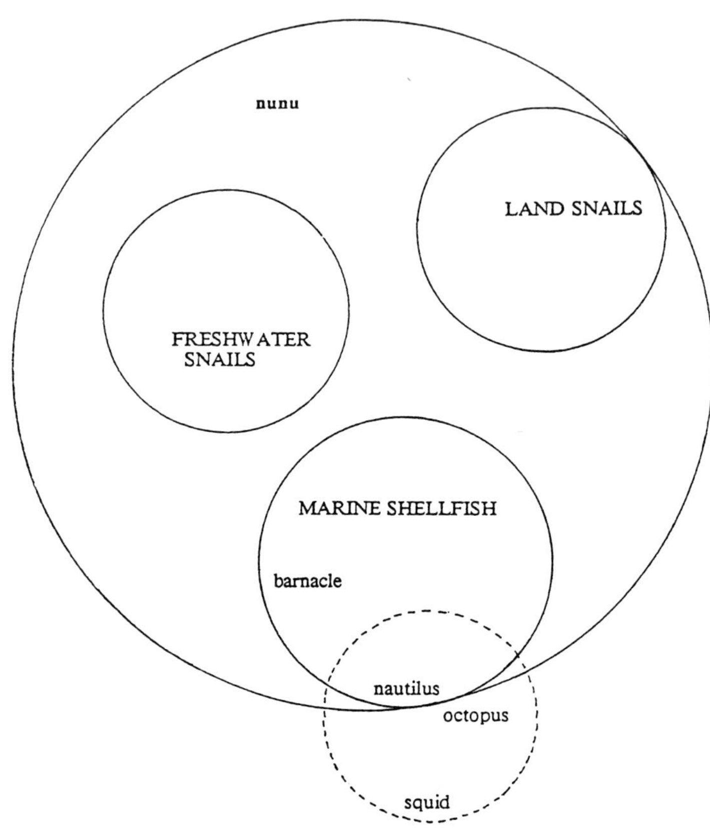

betel lime is discussed in greater detail in Ellen, 1991.

TABLE 20 The social and economic use of molluscs among the Nuaulu.

Species	Eaten	Used for betel lime	Notes
Conus lividus		+	
Conus pertusus			
Conus monachus			
Trochus tenebrica			Occasionally sold. Traditionally used for the manufacture of bracelets and ornaments outside the Nuaulu area (see 11.2.2)
Trochus niloticus	+	+	As above
Trochus flammulatus	+		As above
Trochus costatus	+	+	
Turbo setosus	+	+	
Turbo argyrostomus			
Turbo porophyrites		+	
Lunella coronatus			
Angaria lacinatus			
Drupa morum		+	
Drupa spathulifera		+	
Murex tribulus			
Septaria sanguisaga	+	+	
Septaria sp.			
Cypraea arabica			
Cypraea carneola			
Cypraea lynx	+	+	Poisonous; must be drained and cooked with care
Clithon angulosa			
Clithon subsulcata	+		
Latirus turritus			
Barbatia fusca			
Thiara amarula	+		
Pisania crenilabrum			
Chloritis mima	+	+	Rarely eaten. Sometimes used in the manufacture of lime
Chloritis ungulina	+	+	As above

Nania aulica	+	+	As above
Nanina citrina	+	+	As above
Neritodryas cornea		+	
Nerita antiquata			
Amphidromus			
Melanoides granifera			
Melanoides punctata	+		
Melania hastula	+	+	
Tridacnes elongata	+	+	
Tridacna maxima	+	+	Containers, scrapers, spoons
Periglypta reticulata	+	+	Rarely eaten. Sometimes used in the manufacture of lime. Also used as scrapers, spoons and containers
Nautilus			Used for decorating head-dresses and shields
Squid		+	Only occasionally eaten
Octopus			Rarely seen

Note. I have provided positive entries in columns 2 and 3 only where the indicated usages have been confirmed. It is highly likely that the total number of entries in each case understates the number of species used.

CHAPTER TWELVE

CRUSTACEA

12.1. The crustacean fauna of south central Seram

The macro-crustacean fauna of the central Moluccas (particularly marine fauna) is rich, and has been well-documented since Rumphius first described the main forms. Some indication of the diversity can be obtained by referring to Holthuis van Bentham [Holthuis van Bentham, 1959]. My own field collection and list of certain identifications is slight by comparison. A checklist of crustacea recorded in the Nuaulu region of south central Seram is set out in table 21.

12.2. Nuaulu categories applied to crustacea

Barnacles (THORACIDA) are related in Nuaulu thought with molluscs rather than with crustacea, and are therefore considered elsewhere. Wood lice (**utu wesie, imarua**) seem not to be incorporated into a more inclusive folk category, and certainly show no affinities with other crustacea. The term **utu** suggests some perceived resemblance to body lice (chapter 10.2.8), though no informant articulated this connection. The remaining macro-crustacea are grouped into two large categories: **mitane** (prawns) and **katanopune** (most crabs), with three related forms: **pepeuro** (flat lobsters), **suto** (the rajungan crab, *Portunus pelagicus*) and **kumake** (hermit crabs).

12.2.1 mitane, okote

These terms, which are not further reducible semantically, are applied to prawns, lobsters, shrimps and crayfish, forms generically glossed in AM as 'udang'. **Mitane** is a term which may be used among members of the same sex, but is prohibited in the company of members of the opposite sex, as a short extract from my fieldnotes indicates:

12.6.70 I was taking down details on a diet sheet this evening with Napwai who chuckled when he came to the term **mitane**. The reason for this is apparently that **mitane** also refers to the female genitalia. Among members of the same sex use of this term is permitted. However, in large groups and in the company of members of the opposite sex it is prohibited. Instead, **okote** is used. **Okote** appears to be an archaic Nuaulu term for prawn.

A similar restriction is placed on the term **mara kokowe** (chapter 2.2.1).

Mitane are divided into two basic intermediate categories: **mita nuae** ('sea prawns') and **mita waene** ('freshwater prawns').

12.2.1.2 mita sanane

Sanane = 'waterfall'. This black freshwater prawn is found underneath waterfalls in small streams.

12.2.1.3 mita hanapakue

Hana- = 'hand, arm'; **paku** = 'pin, peg'. The name for this small freshwater prawn may be synonymous with 'finger'.

12.2.1.4 mita sepa

Sepa is the Muslim domain and 'desa' of the same name, prominent in Nuaulu thought, history and political relations. This small freshwater prawn has short (second) antennae.

12.2.1.5 mita uoane, mita pina

Uoane = 'rain'; **pina** = 'female'. These terms are applied to the male and female forms respectively. The (second) antennae are robust, but shorter in the female. Applied to specimens of the genus *Macrobrachium*.

12.2.1.6 mita hahu

Hahu = 'pig'. Very small prawn, with a striped black and yellow body.

12.2.1.7 mita putie

Putie = 'white'. Small freshwater prawn of the genus *Macrobrachium*, possibly *esculentum, javanicum* or *lar*.

12.2.1.1 mita waene

Waene = 'freshwater'. As well as designating a broad general category, **waene** is also applied terminally to a number of species of freshwater prawn, particularly those found in the larger rivers. This term is not used residually; rather it is used systematically for prawns which are fished. It is possible that terms 12.2.1.2 to 12.2.1.6 are regarded as particularly salient, in a sense other than being economically important or widely distributed. The category **mita waene** includes the giant freshwater prawn *Macrobrachium australe* and *Caridina nilotica*.

TABLE 21 Checklist of Crustacea recorded in the Nuaulu region of south central Seram, 1970-1975.

Species	Ecological zones			Nuaulu gloss
	1	2	3	
THORACIDA -barnacles	+	-	-	[chapter 11.2.32]
DECAPODIDA				
NATANTIA -shrimps and prawns[1]				mitane
Generic term for freshwater prawns e.g. *Caridina nilotica*	-	+	-	mita waene
Various freshwater prawns of the	-	+	-	mita sanane
families Palaemonidae, Atyidae	-	+	-	mita hanapakue
and Parastacidae	-	+	-	mita sepa
Macrobrachium australe	-	+	-	mita oane
Macrobrachium australe	-	+	-	mita pina
Macrobrachium sp.	-	+	-	mita putie
poss. *esculentum, lar*				
or *javanicum*				
Penaeidae				
Penaeus monodon,				
Metapenaeus sp.	+	-	-	mita hahu
	+	-	-	mita nuae
NEPHROPSIDEA -lobsters[2]				
-flat lobster				
Scyllaridae				
Thenus orientalis	+	-	-	pepeuro
Nephropidae -spiny lobster				
Panulirus versicolor				
PAGURIDEA -hermit crabs[3]	+	-	-	kumake
	+	-	-	kuma kinihane, kuma moti, kuma hihikuro etc.
BRACHYURA -true crabs[4]				katanopune
-rajungan				
Portunidae -swimming crabs				
Portunus pelagicus	+	-	-	suto
Portunus sanguinolentus				
Ocypodidae				
Ocypode cordimana	+	-	-	katanopu nuae
Ocypode cordimana	+	-	-	katanopu kakante

	Zone 1	Zone 2	Zone 3	
Grapsidae -rock crabs				
Sesarma sp.	-	-	+	**katanopu ai ukune**
Sesarmops impressum	-	+	-	**katanopu nahine**
Ptychognathus riedelii	-	+	-	**katanopu huse**
Metasesarma aubryi				
	-	+	+	**katanopu tanane**
				katanopu kukurisa
	-	-	+	**katanopu uri**
				katanopu sirisa
Potamidae -freshwater crabs				
	-	+	-	**katanopu tonate unte**
	-	+	-	**katanopu putie**
	-	-	+	**katanopu makapotae iane**
Gecarcinidae -land crabs				
Poss. *Cardiosoma* sp.				**katanopu manapesi**
Birgus latro -the coconut crab	-	-	+	**katanopu sipu-sipu**
				katanopu nosue
ISOPODA -wood lice				
Thrombium sp.	-	-	+	**utu wesie, imarua**

Key. Zone 1 = marine; zone 2 = freshwater; zone 3 = terrestrial. *Notes.* 1. Suborder. 2. Superfamily. 3-4. Section.

12.2.1.8 mita nuae

Nuae = 'sea'. As well as designating a broad general category, this term is applied terminally to many species of marine prawn.

12.2.2 pepeuro

This term, which is not further reducible semantically, is applied to the flat lobster *Thenus orientalis* and possibly also *Panulirus versicolor*.

12.2.3 kumake

Not further reducible semantically. The term is applied to hermit crabs. Rumphius notes that on Ambon 'there are people who pretend that certain kinds of hermit crabs, having grown so large that they can no longer find suitable shells, stroll about nude, finally metamorphizing into coconut crabs (*Birgus latro*)' [quoted in Holthuis 1959: 189]. He thought this improbable. The same belief is held by the Nuaulu, and discussed in chapter 6.3 of *The*

Cultural Relations of Classification.

Most hermit crabs are binomialised by adding the name of the shell which they inhabit. Thus we have **kuma kinihane, kuma hihikuro, kuma mata putie, kuma tapako** and so on; and chapter 11 should be consulted for the identification and meaning of these terms. Obviously, such categories bear little resemblance to phylogenetic taxa as one hermit crab may occupy numerous, successively larger, shells in the course of its life. One term, however, does not follow this pattern:

12.2.3.1 kuma moti

Moti = 'rocks and pools exposed at low tide'. A small hermit crab. The term may be employed loosely to cut across categories defined in terms of shell type.

12.2.4 suto

A term applied to the large edible rajungan crab, *Portunus pelagicus* (and perhaps also *P. sanguinolentus*), sometimes described in AM as 'koli-koli', meaning also a large dugout canoe. This is the only term for a crab never prefixed by **katanopu**, and appears to be regarded as a quite distinct category.

12.2.5 katanopune

The term is applied to most true crabs.

12.2.5.1 katanopu tanane

Tanane ='land'. The term is applied to a large red freshwater crab, about 5 cm by 10 cm. Prob. = **tanae** (R.B.).

12.2.5.2 katanopu kukurisa

The meaning of **kukurisa** is unclear. This crab is related to **katanopu tanane** although the carapace is slightly different.

12.2.5.3 katanopu uri

Uri = 'banana'. Found on banana and plantain leaves.

12.2.5.4 katanopu sirisa

The meaning of **sirisa** is unclear. This naturally red crab turns a much brighter hue on cooking, and has a width of about 5 cm.

12.2.5.5 katanopu makapotae iane

Lit. 'unripe kenari eating crab'; **iane** ('kenari' in AM) = nut of *Canarium commune*. Red and brown freshwater species; 5 cm wide.

12.2.5.6 katanopu nahine

White freshwater crab living on river banks; certainly includes *Sesarmops impressum*.

12.2.5.7 katanopu huse

Huse may be a contraction of **husue, husui,** 'boil, inflamed tumour'. Freshwater; 5 cm wide; certainly includes *Ptychognathus riedelii*.

12.2.5.8 katanopu manapesi

Lit. 'bearded crab', referring to 'hair' on the legs and 'breast' of this 5 cm wide crab. Poss. *Cardiosoma* sp.

12.2.5.9 katanopu sipu-sipu

Meaning unclear. Small edible crab; found on coconut palms to landward side of village of Rohua. Possibly coconut crab, *Birgus latro*.

12.2.5.10 katanopu ai ukune

Ai ukune = 'treetop', far forest.

12.2.5.11 katanopu tonate unte

Onate = 'large', **unte** = 'skin'. Large black, edible crab; yellow and black underside. Found on bed of larger rivers, such as the Upa. Poss. *Cardiosoma* sp.

12.2.5.12 katanopu putie

Putie = 'white'. Largish white crab found in rivers. Regarded as quite inedible.

12.2.5.13 katanopu nosue

Nosue or **nosui** means 'in an unwrapped state, naked'. The term may refer to crabs which have just shed an old carapace. Applied to a soft black crab, about 5 cm in width.

12.2.5.14 katanopu nuae

Nuae = 'sea'. Residual term applied to all marine crabs not otherwise distinguished.

12.2.5.15 katanopu manante, katanopu kakante

I have heard both terms on different occasions, and they are probably free variants. Refers to a crab with large orange claws, orange carapace and pink legs. On **kakante** see chapter 11.2.28. Applied to *Ocypode cordimana*.

12.2.6 utu wesie, imarua

All woodlice, including *Thrombium*.

12.3 The social and economic uses of crustaceans

None of the prawns are used for the making of 'terasi', the fermented fish paste important in Javanese, and also in some Ambonese, cooking. The species used for this purpose include *Penaeus monodon, P. indicus* and *Heterpenaeus* spp. Larger edible freshwater prawns are regularly collected from along the banks of streams and larger rivers [Ellen 1993: plate 1.5); marine crustacea, especially crabs, are rarely eaten.

CHAPTER THIRTEEN

ARACHNIDS

13.1 The arachnid fauna of south central Seram

As with insects, only a very few specimens compared with the total number of known species were collected in the field. Nevertheless, they do cover most species commonly encountered and named by the Nuaulu. A checklist of arachnid specimens recorded in the Nuaulu area during field-work is presented in table 22.

13.2 Nuaulu categories applied to arachnids excepting Acarida

13.2.1 kanopone

Scorpions (SCORPIONIDA) and possibly also whip-scorpions (*URO-PYGI*).

13.2.2 riko-riko, nau asue

The first term is consistently applied to harvestmen (*PHALANGIDA*). In the second, **nau** is a general term for augury and divination; **asu** = 'dog', **asu-** = 'cheek' + non-human possessive pronominal suffix. **Nau asue** is applied to the harvestman *Altobunus formosus.* It is probably a synonym for **riko-riko,** being used as a nick-name in circumstances in which it seems auspicious.

13.2.3 kahuneke hatu nohu inae

Hatu nohu(e), meaning 'cavernous rock outcrop, cave', indicates the habitat of this spider; **inae** = 'mother'. As **kahunekete** is the generic term for spider we thus have 'mother cave spider'. Applied quite specifically to tailess whip scorpions, and generally encountered in rock fissures when hunting bats.

13.2.4 kahuneke ai ukune

Ai ukune = 'treetop', far forest. Applied to various kinds of long-bodied spider, including *Theridion* and possibly *Nephilia.* It therefore seems to be applied to both hunters and spinners of irregular webs in forest habitats.

13.2.5 kahuneke titie

Titie = 'hot', so-called on account of the ability of this round-bodied spider to bite humans and cause a painful swelling. Applied to *Argiope*, and perhaps *Heteropoda;* also applied to the local variety of the Black Widow, *Latrodectus*, and members of the families Clenidae, Clubionidae and Theraphosidae. Commonly found around human habitations and in cleared areas.

13.2.6 kahuneke metene

Metene = 'black'. Almost always applied to long-bodied spiders, such as *Nephilia*.

13.2.7 kahuneke masikune

Masikune = 'yellow'. Not applied to any specimen with certainty, but sometimes used to describe *Argiope* with yellow markings.

13.2.8 kahuneke nikate

Nikate = 'pattern, drawing, design': refers to orb-web or markings of *Argiope,* and perhaps other spiders (e.g. Salticidae).

13.2.9 kahuneke numa

Numa = 'house'. Applied to the most common domestic jumping spider (*Heteropoda venatoria*), found living on the inner and outer walls of houses and other structures. This spider has a benign reputation among the Nuaulu for its depredations on cockroaches and other undesirable insects. Agelenidae may be present, but probably also includes members of the following families: Heteropodidae, Psechridae, Pholcidae and Argiopidae.

13.2.10 kahuneke onate

Onate = 'large'. I have only come across this term being applied to long-bodied spiders described by other informants as **kahuneke wala-wala,** though R.B. reports that the term also appears to be used interchangeably with **kahuneke masikune**, indicating the inclusion of *Argiope*.

13.2.11 (kahuneke) wala-wala

Wala-wala = 'web'. Applied to long-bodied spiders including *Nephilia;* to all web-spinners. In fact, web-spinning spiders are sometimes simply called **wala-wala** (the product coinciding terminologically with the producer). For a long time this usage was a source of perplexity to me.

TABLE 22 Checklist of Arachnid specimens recorded in the Nuaulu region of south central Seram, 1970-1975.

Species	Ecological zones			Nuaulu gloss
	1	2	3	
SCORPIONIDA -scorpions				
Isometrus maculatus	+	+	+	**kanopone**
UROPYGIDA				
UROPYGI -whip scorpions				
Uropygi sp.				
AMBYPYGI -tailess whip scorpions				
Charontidae	-	-	+	**kahuneke hatu nohu inae**
ACARIDA -mites and ticks				
Aponomma gervaisi				
Seiulus amboinensis				[chapter 10.2.7, 10.4]
PHALANGIDA -harvestmen				
Altobunus formosus	+	+	+	**riko-riko, nau asue**
ARANEIDA -spiders (hunters)				
Sparassidae				
Nephilia sp.	-	+	+	**kahuneke wala-wala, kahuneke wala-wala sonu, kahuneke metene, kahuneke onate, kahuneke ai ukune**
Nephilia maculata				
Heteropoda venatoria	+	-	-	**kahuneke numa, kahuneke titie**
(web-spinners)				
Theridiidae -irregular webs				
Theridion sp.	+	+	+	**kahuneke ai ukune**
Latrodectus hasselti -'black widow'	+	+	-	**kahuneke titie**
Argiopidae -orb webs				
Argiope sp.	+	+	+	
Argiope aemula	+	+	+	
				kahuneke titie
				kahuneke nikate, kahuneke masikune

Key. Zone 1 = village; zone 2 = cultivated areas; zone 3 = forest.

13.2.12 kahuneke wala-wala sonu

Sonu refers to **ikae sonu,** needlefish of the genera *Strongylura* and *Tylosurus.* The web of this round-bodied spider is collected, twisted together and used as bait (or more correctly as a substitute for a hook) when line-fishing for needlefish, who find it difficult to extricate their teeth from the compressed webby substances (chapter 9.2.12). Only two individuals in Rohua (Naunepe and Komisi) claim to use this technique, and it has almost certainly been acquired from outsiders.

13.3 Variation, arrangement of categories and the social significance of arachnids

All ARANEIDA, together with cave scorpion spiders (ABLYPYGI), are classified as **kahunekete.** This is terminologically expressed in every name collected except **wala-wala,** where the prefix **kahuneke** is optionally dropped. It is clear that the various terms for types of spider and the content of the categories which they label do not easily yield a taxonomic representation. The phylogenetic content of particular categories may be wide-ranging and overlapping, reflecting a series of asymmetric cross-cutting contrasts between marked categories and their residues: *large : (other), house-dwelling : (other), black : (other), stinging : (other).*

Harvestmen (PHALANGIDA) are regarded by many Nuaulu as related (on morphological grounds) to spiders, and some people actually suggested that they might be a kind of **kahunekete.** Scorpions are regarded as unaffiliated. Aquatic arachnids (MEROSTOMATA) are peripherally linked (in a residual way) with other kinds of marine invertebrates and are therefore considered in chapter 14. As can be seen from an inspection of table 13.1, the combination of functional and morphological characteristics used to generate particular names (body shape, colour, web type, sting, habitat) result in a rather uneasy fit between phylogenetic and Nuaulu classification, and with the exception of **kahuneke hatu nohu inae** are probably applied rather loosely.

The term **kahunekete** resembles **kau** (the contracted form of **kauke,** as also applied to crickets, grasshoppers and locusts) plus the root **neke** (= 'sleep'). It may not be too fanciful to read into this the representation of a classificatory linkage. The general similarities between arachnids and insects are fairly obvious: they are small, invertebrate, have hard exoskeletons, have six or eight legs and scamper; **neke** would appear to relate to web-spinning and the habit of sitting motionless for long periods at a stretch. If this is so, then a characteristic of prominent members of a category have been appropriated to label the category as a whole. This is a practice which we are most familiar with in respect to birds, though it is an inevitable consequence of any classification which relies on salient distinctive features.

The only mundane technical practice connected with arachnids is the use of the web of **kahuneke wala-wala sonu** for catching needlefish. As a group, arachnids are not feared. Some are known to be harmful to humans **(kahuneke titie, kanopone)**; but others may be auspicious (**nau asue**). The only general belief concerning spiders which I have come across, and this is associated with web-spinners, is that they 'lie to dogs'. On one occasion in 1975 I was returning to Rohua with Kaiisa when his dogs began whinning and snapping and walking around in circles. I asked why this should be and was told that webs stretching across the path had led the dogs to lose their way, and that they would now find an alternative route home. The belief is that this is wilful interference on the part of spiders, but nevertheless playful and non-malevolent.

CHAPTER FOURTEEN

WORMS, MYRIAPODS, ECHINODERMS
AND OTHER RESIDUAL FORMS

The contents of this final chapter are admittedly residual. It covers all those forms which the Nuaulu would recognise as animals, but which so far have not been considered. It includes representatives from the following phyla: ANNELIDA, PLATYHELMINTHES, NEMATODA, ECHINODER-MATA and, from among the ARTHROPODA, *CHILOPODA, DIPLOPODA* and *MEROSTOMATA*.

No attempt has been made to ascertain the total numbers of species on Seram for these groups, but as with insects and arachnids, only a very few specimens compared with the total number of known species were collected in the field. Nevertheless, they contain all forms commonly encountered and named by the Nuaulu. A checklist of the major forms recognised among these groups and of actual specimens collected in the field is presented in table 23.

14.1 Nuaulu terms applied to worms, myriapods, echinoderms and other residual forms

14.1.1-4 susue, sohane, nikonake, mnatone

Worms in the broadest sense represent a covert group which includes platyhelminthes, nematodes and annelids. Endoparasitic worms are seen as the larvae of certain insects, a belief which must be said to constitute a Nuaulu theory of infection. Intestinal nematodes (**susue, sohane**) are regarded as the metamorphized larvae of fruit-flies (**mumna**) and other flies (**sohane inae** : chapter 17.2.45), which are thought to enter the body through wounds. The distended stomachs of malnourished children are said to be full of **sohane.** The tapeworm (**nikonake:** *Taenia* spp.) is also judged to be the product of such a metamorphosis, but ideas tend to be less clear-cut. **Mnatone** are sea-worms, including the well-known palalo worm (*Eunice viridis*). This is harvested annually in large numbers in certain Moluccan villages, such as at Latuhalet on Ambon island, where it is known as 'lawer'.

14.1.5 ai ntone

Lit. 'tree sap'. Large black and white perichaet earthworm which lives in decaying trees, particularly dry bamboo and sago leaf petiole. Alleged to have a dangerous bite, and to be able to enter bodily orifices.

14.1.6 tuaman (i)nae, tua nanae (R.B.)

Lit. 'mother of the earth' and 'child of the earth', respectively. With the exception of those species assigned to the category **ai ntone,** all of the earthworm species encountered by the Nuaulu (and there are perhaps upwards of 746 known from Seram) are designated by this term (e.g. *Pheratima, Pontoscolex* spp.).

14.1.7 sanina, sanna, inae (R.B.)

Applied to all terrestrial and freshwater leeches. These are creatures of the damp forests, particularly of the sago swamp forest, but I have always been surprised by their infrequency on Seram, compared with what one would believe from the standard travellers tales of Borneo. They are rarely encountered in the immediate vicinity of the village, but it is said that if you get a leech in your eye, it may swell up to such an extent as to be a cause of death.

14.1.8 niniane

Centipedes, known in AM as 'kakisaribu'. **Niniane** may be roasted and eaten. Three kinds are recognised. In each case the prefixes **nini** and **ninia** (reported by R.B.) appear to be in free variation.

14.1.8.1 ninia oni, nin ikine

The meaning of **oni** is not known, but it is clearly the same suffix as in **imanine on,** applied to a type of small red wasp (chapter 10.2.31.2). As it is applied to smaller species of *Scolopendra (morsitans),* it appears to refer (as with the wasp) at least in part to their relative smallness. **Nini oni** can give a nasty bite, which may occasionally prove fatal. One informant recalled a kinsman who died within 48 hours of being bitten.

14.1.8.2 ninia bunara

If **oni** is an allusion to smallness, then **bunara** must be an allusion to bigness (c.f. **imanine bunara;** chapter 10.2.31.3). **Bunara** is a toponym for the Nuaulu village and river of the same name (figure 3). Applied to large specimens of *Scolopendra* (e.g. *subspinipes*).

TABLE 23 Checklist of annelids, echinoderms, myriapods and related forms featuring in Nuaulu terminology and knowledge.

Species	Ecological zones				Nuaulu gloss
	1	2	3	4	
COELENTERATA					
- jellyfish	-	-	+	-	**nuae huae**
PLATYHELMINTHES - flatworms					
Taenia - tapeworm	-	-	-	+	**nikonake**
NEMATODA -roundworms					
human intestinal parasites	-	-	-	+	**susue, sohane**
incl. *Eunice viridis*	-	-	+	-	**mnatone**
ANNELIDA					
POLYCHAETA	-	-	+	-	**mnatone**
OLIGOCHAETA -earthworms					**tuaman (i)nae**
Pheretima (Pheretima) ceramensis	+	-	-	-	
Pontoscolex corethrurus	+	-	-	-	
PERICHAETA	+	-	-	-	**ai ntone**
HIRUDINEA -leeches	+	-	-	-	**sanna**
CHILOPODA -centipedes					**niniane**
Scolopendra morsitans	+	-	-	-	**nini oni**
Scolopendra subspinipes	+	-	-	-	**nini Bunara**
	+	-	-	-	**nini anane**
DIPLOPODA -millipedes					**nikenuke**
Thyropygus -giant millipede	+	-	-	-	
Rhinocricus -giant millipede	+	-	-	-	**nike putie, nike metene, nike msinae**
-other millipedes					
MEROSTOMATA -horseshoe crabs					
Limulus moluccensis	-	-	+	-	**mimi**
ECHINODERMATA					
ASTEROIDEA, OPHIUROIDEA	-	-	+	-	**une nuae**
- starfish					
different colour phases of e.g.					
Protoreaster nodosus					
Pentaceraster					**une msinae, une masikune, une marae, une putie, une metene**
prob. incl. *Linckia laevigata*					**une marae**

ECHINOIDEA -sea urchin	-	-	+	-	**tinene**
prob. *Echinotrix diadema*	-	-	+	-	**tine metene**
prob. *Tripneusteus gratilla*	-	-	+	-	**tine metene**
prob. *Diadema setosum*	-	-	+	-	**tine msinae**
prob. *Diadema saxatile*	-	-	+	-	**tine msinae**
HOLOTHUROIDEA -sea cucumbers, trepang	-	-	+	-	**taripan**

Key. Zone 1 = terrestrial; zone 2 = freshwater; zone 3 = marine (including intertidal regions); zone 4 = endoparasitic.

14.1.8.3 nini anane

Elicited by R.B. Meaning and reference unclear.

14.1.9 nikenuke

Glossed in AM as 'lekar-lekar' or 'keluwing', and applied by the Nuaulu to all kinds of millipedes. Three kinds are recognised, though not by all adults, and the terms are rarely used in ordinary discourse. In each case the prefixes **nike** and **nikenu** (reported by R.B.) appear to be in free variation.

14.1.9.1 nike putie

Putie = 'white'.

14.1.9.2 nike metene

Metene = 'black'.

14.1.9.3 nike msinae

Msinae = 'red'.

14.1.10 tinene

Sea-urchins. Two types are recognised, but all *ECHINOIDEA* are rare along that part of the south Seramese coast with which the Nuaulu are most familiar.

14.1.10.1 tine metene

Metene = 'black'; Probably includes *Echinotrix diadema* and perhaps also *Tripneusteus gratilla,* the eggs of which may be eaten.

14.1.10.2 tine msinae

Msinae = 'red'; Probably includes *Diadema setosum* and perhaps *Diadema saxatile.*

14.1.11 une nuae

Nuae is 'sea', and **une** (generally in the reduplicated form **une-une**) is translatable as caterpillar (chapter 10.3). In this context, however, the latter term appears to be a corruption of **one,** meaning 'star', or more remotely **uhune,** 'flower bud'. The first was the only folk-etymology proferred, and is perhaps the most plausible. The AM gloss is 'bintang laut', also meaning 'sea star', and the Nuaulu may be a translation of this or a similar term in Sepa. Five kinds of **une nuae** appear to be recognised, the first four being applied to the different colour phases of *Protoreaster nodosus,* or different colour varieties of *Protoreaster;* although by extension either the terminal labels or the undifferentiated term **une nuae** may be applied to other seastars which are occasionally encountered, such as *Pentaceraster* or *Protoreaster.* I have never known the Nuaulu to eat seastars, or use them for other purposes, as is the case among other coastal peoples of Seram.

14.1.11.1 une msinae
Msinae = 'red'.

14.1.11.2 une masikune
Masikune = 'yellow'.

14.1.11.3 une putie
Putie = 'white'.

14.1.11.4 une metene
Metene = 'black'. Elicited by R.B.

14.1.11.5 une marae
Marae = 'blue-green'; applied to species such as *Linckia laevigata.*

14.1.12 taripan

Possibly derived from AM, and applied to the many varieties of seacucumber, generally known nowadays in AM as 'trepang'. Not common in the inland waters off the south coast of Seram between Elpaputih and Teluti Bay.

14.1.13 nuae huae

Lit. 'fruit of the sea': jellyfish of all kinds.

14.1.14 mimi

Cognate with AM 'mimi', a term which may also be present in various other Central Maluku languages. This horseshoe crab (*Limulus moluccanus*) lives in shallow water along sandy and muddy shores, spending most of its time burrowing; not frequently encountered.

14.2 Uses and arrangement of categories

Only sea-urchins in this diverse assortment of invertebrates can be said to in any way constitute a generally recognised source of food, and these are only occasionally eaten. This is partly on account of their local scarcity and partly because the Nuaulu have other historically determined food preferences in the realm of marginal foods. Indeed, the general unwillingness to eat marine invertebrates other than molluscs, and a very rudimentary knowledge of their anatomy and habits, must be directly connected with the inland origins of the Nuaulu. Giant centipedes are sometimes roasted and eaten.

I have noted (14.1.1-3) that worms constitute a covert grouping, with partial connections with certain insects. Additionally, worms, centipedes and millipedes, have classificatory links with snakes, which I have discussed in an earlier chapter (7.3). Echinoderms and jellyfish do not in themselves form a covert group, but they are contrasted with all other forms considered here - with the possible exception of marine worms - by their marine lifestyle. In this sense, their closest links are with the category **ikae** (fish).

LIST OF ABBREVIATIONS

alt	alternatively
AM	Ambonese Malay
arch	archaic
BM	British Museum (Museum of Mankind), London
CM	Central Maluku
cm	centimeters
D	diameter
H	height
incl	including
Ind	Indonesian
indet	indeterminate
kg	kilograms
L	length
lit	literally
m	meters
n	noun
nr	near
PAN	Proto Austronesian
PCM	Proto Central Maluku
prep	preposition
prob	probably
RB	Rosemary Bolton; in acknowledgement of the source of a term or meaning
sp (spp)	species (plural)
♀	female
♂	male
?	uncertain determination
0	in a diagram indicates a covert category

Reference to specimens in museums and private collections is indicated using round brackets as in the following example: (e.g. BM As. 1.177 and Ellen 1970.617).

LIST OF REFERENCES

Amir, H. and J. Wind 1978: The nature reserves of Way Mual and Way Nua, central Seram, Maluku. *Field Report* **8**, FO/JNS/73/013.

Anon. 1908: Gouden beeldje (Çiva) van Amahei. *Notulen van de Algemeene en Bestuursvergaderingen van het Bataviaasch Genootschap van Kunsten en Wetenschappen* **46**, 67, lxi.

Barnes, R. 1989: *The ikat textiles of Lamalera: a study of an eastern Indonesian weaving tradition* (Studies in South Asian Culture **14**). Leiden: E. J. Brill.

Bemmel, A.C.V. van 1948: A faunal list of the birds of the Moluccan islands. *Treubia* **19**, 323-401.

Bemmel, A.C.V. van and K. H. Voous 1953: Supplement to the faunal list of the birds of the Moluccan Islands. *Beaufortia* **4**, 32 1-7.

Benthem Jutting, W. S. S. van 1959: Rumphius and malacology. In *Rumphius memorial volume.* (ed.) de Wit, H. C. D. Baarn: Hollandia N. V.

Berlin, B., D. Breedlove and P. Raven 1974: *Principles of Tzeltal plant classification : an introduction to the botanical ethnography of a Mayan-speaking people of the highland Chiapas.* New York: Academic Press.

Bowler, J. and J. Taylor 1989: An annotated checklist of the birds of the Manusela National Park, Seram (birds recorded on the Operation Raleigh expedition). *Kukila* **4**, (1-2), 3-29.

Brown, C. H. 1984: *Language and living things: uniformities in folk classification and naming.* New Brunswick, New Jersey: Rutgers University Press.

Bulmer, R. N. H. and M. Tyler 1968: Karam classification of frogs. *Journal of the Polynesian Society* **77**, 333-85.

Bulmer, R. 1968: Worms that croak and other mysteries of Karam natural history. *Mankind* **6**, 621-639.

Bulmer, R. 1969: *Field methods in ethno-zoology with special reference to the New Guinea highlands* (Mimeographed). Port Moresby: University of Papua and New Guinea.

Bulmer, R. N. H. and J. I. Menzies 1972-3: Karam classification of marsupials and rodents. *Journal of the Polynesian Society* **81-82**, 1,4 472-499, 86-107.

Bulmer, R. N. H. and J. I. Menzies 1972-3: Karam classification of marsupials and rodents. *Journal of the Polynesian Society* **81-82**, (1), 472-

499, (4), 86-107.

Bulmer, R. N. H. 1974: Memoirs of a small game hunter: on the track of unknown animal categories in New Guinea. *Journal d'Agriculture tropicale et de Botanique appliqée* **21**, 79-99.

Bulmer, R. N. H., J. I. Menzies and F. Parker 1975: Kalam classification of reptiles and fishes. *Journal of the Polynesian Society* **3**, 267-308.

Burkill, T. H. 1935: *A dictionary of the economic products of the Malay peninsula* (2 volumes). London: Crown Agents for the Colonies.

Cooley, F. L. 1962: *Ambonese adat: a general description* (Cultural Report Series **10**). New Haven, Conn: Yale University Southeast Asia Studies.

Darlington, P. J. 1957: *Zoogeography: the geographical distribution of animals.* New York: Wiley.

Diakonoff, A. 1959: Rumphius as an entomologist. In *Rumphius memorial volume.* (ed.) de Wit, H. C. D. Baarn: Hollandia N. V.

Ekris, A. van 1864-5: Woordenlijst van eenige dialecten der landtaal op de Ambonische eilanden. *Mededeelingen van wege het Nederlandsch Zendeling-genootschap* **9**, 109-136.

Ellen, R. F. 1972: The marsupial in Nuaulu ritual behaviour. *Man* **7**, 223-38.

Ellen, R. F. 1973: *Nuaulu settlement and ecology: an approach to the environmental relations of an eastern Indonesian community.* Unpublished thesis submitted for the degree of Doctor of Philosophy in the University of London.

Ellen, R. F. 1975: Variable constructs in Nuaulu zoological classification. *Social Science Information* **14**, 201-28.

Ellen, R. F. 1975: Non-domesticated resources in Nuaulu ecological relations. *Social Science Information* **14**, 5 51-61.

Ellen, R. F., A. F. Stimson and J. Menzies 1976: The content of categories and experience. The case for some Nuaulu reptiles. *Journal d'Agriculture Tropicale et de Botanique Appliquée* **24**, 3-22.

Ellen, R. F., A. F. Stimson and J. Menzies 1976: Structure and inconsistency in Nuaulu categories for amphibians. *Journal d'Agriculture Tropicale et de Botanique Appliquée* **23**, 125-38.

Ellen, R. F. 1978: The contribution of H.O. Forbes to Indonesian ethnography: a biographical and bibliographical note. *Archipel* **15**, 135-159.

Ellen, R. F. 1978: Restricted faunas and ethnozoological inventories in Wallacea. In *Man and nature in Southeast Asia.* (ed.) Stott, P. H. London: School of Oriental and African Studies.

Ellen, R. F. 1978: *Nuaulu settlement and ecology: an approach to the environmental relations of an eastern Indonesian community* (Verhandelingen van het Koninklijk Instituut voor Taal-, Land- en Volkenkunde 83). The Hague: Martinus Nijhoff.

Ellen, R. F. 1983: Semantic anarchy and ordered social practice in Nuaulu personal naming. *Bijdragen tot de Taal-, Land- en Volkenkunde* **139**, 18-45.

Ellen, R. F. 1984: Trade, environment and the reproduction of local systems in the Moluccas. In *The Ecosystem Concept in Anthropology* (AAAS Selected Symposium **92**). 163-204, (ed.) Moran, E. F. Boulder CO: American Association for the Advancement of Science.

Ellen, R. F. 1985: Patterns of indigenous timber extraction from Moluccan rain forest fringes. *Journal of Biogeography* **12**, 559-587.

Ellen, R. F. 1986: Microcosm, macrocosm and the Nuaulu house: concerning the reductionist fallacy as applied to metaphorical levels. *Bijdragen tot de Taal-, Land- en Volkenkunde* **142**, (1) 1-30.

Ellen, R. F. 1988: Foraging, starch extraction and the sedentary lifestyle in the lowland rainforest of central Seram. In *Hunters and gatherers 1: history, evolution and social change.* (eds.) Ingold, T., D. Riches and J. Woodburn. London: Berg.

Ellen, R. F. 1988: Ritual, identity and the management of interethnic relations on Seram. In *Time past, time present, time future: essays in honour of P. E. de Josselin de Jong* (Verhandelingen van het Koninklijk Instituut voor Taal-, Land- en Volkenkunde **131**). 117-135, (eds.) Moyer, D. S. and H. J. M. Claessen. Dordrecht-Providence: Foris Publications.

Ellen, R. F. 1991: Nuaulu betel chewing: ethnobotany, technique and cultural significance. *Cakalele* **2** (2), 1-25.

Ellen, R. F. 1993: *The cultural relations of classification: an analysis of Nuaulu animal categories from central Seram.* Cambridge: Cambridge University Press.

Forbes, H. O. 1885: *A naturalist's wanderings in the eastern archipelago: a narrative of travel and exploration from 1878 to 1883.* London: Sampson Low, Marston, Searle and Rivington.

Gathorne-Hardy (Lord Medway), G. 1978: *The wild mammals of Malaya (peninsula Malaysia) and Singapore.* Kuala Lumpur: Oxford University Press.

Gijsels, A. 1871: Grondig verhaal van Amboyna, 1621. *Kroniek van het Historisch Genootschap, Utrecht* **27**, 348-394, 397-444, 450-494.

Gimlette, J. D. 1971 [1915]: *Malay poisons and charm cures.* Kuala Lumpur: Oxford University Press.

Glover, I. C. 1971: Prehistoric research in Timor. In *Aboriginal man and environment in Australia.* (eds.) Mulvaney, D. J. and J. Golson. Canberra: Australian National University Press.

Glover, I. C. 1986: Archaeology in eastern Timor, 1966-67. *Terra Australis, 11*, Canberra: Department of Prehistory, Research School of Pacific Studies, Australian National University .

Gonda, J. 1973: *Sanskrit in Indonesia.* New Delhi: International Academy of Indian Culture.

Greshoff, M. 1902: *Rumphius gedenkboek, 1702-1902.* Haarlem: Koloniaal Museum.

Groves, C. P. 1985: On the agriotypes of domestic cattle and pigs in the Indo-Pacific region. In *Recent advances in Indo-Pacific prehistory.* (eds.) Misra, V. N. and P. Bellwood. Leiden: E. J. Brill.

Heekeren, H. R. van 1958: *The bronze-iron age of Indonesia* (Verhandelingen van het Koninklijk Instituut voor Land-, Taal- en Volkenkunde 22). 's-Gravenhage: Martinus Nijhoff.

Heider, E. R. 1972: Probabilities, sampling and ethnographic methods: the case of Dani colour names. *Man (N. S.)* 7, 3 448-66.

Hembree, E. D.1980: Biological aspects of the Cetacean fishery at Lamalera, Lembata. **1428.**

Holthuis van Bentham, L. B. 1959: Notes on pre-linnean carcinology (including the study of Xiphosura) of the Malay archipelago. In *Rumphius memorial volume.* (ed.) de Wit, H. C. D. Baarn: Hollandia N. V.

Hunn, E. 1975: A measure of the degree of correspondence of folk to scientific biological classification. *American Ethnologist* **2**, (2) 309-27.

Hunn, E. 1976: Toward a perceptual model of folk biological classification. *American Ethnologist* **3**, (3), 508-24.

Hunn, E. 1977: *Tzeltal folk zoology: the classification of discontinuities in nature.* London: Academic Press.

Iskandar, T. 1970: *Kamus dewan.* Kuala Lumpur: Dewan Bahasa dan Pustaka.

Jensen, A. E. 1939: *Hainuwele:Volkserzählungen von der Molukken Insel Ceram* (Ergebnisse der Frobenius Expedition 1937-1938 in die Molukken und nach Holländisch Neu-Guinea 1). Frankfurt am Main: Klostermann.

Kraneveld, F. C. 1959: Rumphius and veterinarian science in the eastern archipelago. In *Rumphius memorial volume.* (ed.) de Wit, H. C. D. Baarn:

Hollandia N. V.

Macdonald, A. A., J. E. Hill, D. J. Kitchener, L. Charleton, Boeadi and R. Cox n.d.: *The mammals of Seram, with notes on their biology and local usage.* Unpublished: Manuscript.

Majnep, I. S. and R. Bulmer 1977: *Birds of my Kalam country.* Auckland: Auckland University Press, Oxford University Press.

Nolthenius, A. B. T. 1935: *Overzicht van de literatuur vetreffende de Molukken, 2 (1922-1933).* Amsterdam: Molukken-Instituut.

Peeters, A. 1979: Nomenclature and classification in Rumphius's 'Herbarium Amboinense'. In *Classifications in their social context.* (eds.) Ellen, R. F. and D. Reason. London: Academic Press.

Pocock, R. I. 1933: The civet cats of Asia. *Journal, Bombay Natural History Society* **26**, 423-449, 629-656.

Rand, A. L. and E. T. Gilliard 1967: *Handbook of New Guinea birds.* London: Weidenfeld and Nicolson.

Ribbe, C. 1889: Betrage zur Lepidopteren-Fauna von Gross-Ceram. *Iris, Deutsche Entomologische Zeitschrift* **2**, 187-265.

Ribbe, C. 1892: Ein Aufenthalt auf Grosz Seram. *Jahresberichte des Vereins fur Erkunde* **22**, 129-216.

Rouffaer, G. P. and H. H. Juynboll 1914: *De batik-kunst in Nederlandsch-Indie en hare geschiedenis op grond van materiaal aanwezig in 's Rijks Etnographisch Museum en andere openbare en particuliere verzamelingen in Nederland.* Haarlem: Kleinmann.

Rubenkoning, J. A. 1959: Preface. In *Rumphius memorial volume.* (ed.) de Wit, H. C. D. Baarn: Hollandia N. V.

Ruinen, W. 1928: *Overzicht van de literatuur betreffende de Molukken (1550-1921).* Amsterdam: Molukken Instituut.

Rumphius, G. E. 1741 [1705]: *D'Amboinsche Rariteitkamer.* Amsterdam: Third Impression.

Sachse, F. J. P. 1907: *Het eiland Seran en zijne bewoners.* Leiden: Brill.

Salvadori, T. 1880-82: Ornitologia della Papuasia e delle Molucche. *Memorie della Regio Accademica della Scienze di Torino* **33** (1-3).

Schafer, E. H. 1967: *The vermilion bird: T'ang images of the south.* Berkeley and Los Angeles: University of California Press.

Siebers, H. C. 1930: Aves. Fauna Buruana. *Treubia* **7**, 165-303.

Statistik Tahunan 1974: *Maluku Dalam Angka.* Ambon: Kantor Sensus dan Statistik Propinsi Maluku.

Stresemann, E. 1914: Die Vogel von Seran. *Novitiates Zoologie* **21**, 25-153.

Stresemann, E. 1927: Die Lauterscheinungen in den amboinischen Sprachen. *Zeitschrift fur Eingeboren-Sprachen* **10**,

Tweedie, M. 1970: *Animals of southern Asia*. London: Hamlyn.

Valentijn, F. 1724-26: *Oud en Nieuw Oost-Indiën*. Dordrecht - Amsterdam: 8 vols.

Wallace, A. R. 1962 (1869): *The Malay Archipelago*. New York: Dover.

Weber, M. and L. F. Beaufort 1913-1922: *The fishes of the Indo-Australian archipelago*. Leiden: Brill.

White, C. M. N. 1973: Night herons in Wallacea. *Bulletin of the British Ornithological Club* **93**, 175-76.

White, C. M. N. 1975: The problem of the cassowary in Seran. *Bulletin of the British Ornithologist's Club* **95**, 4 165-170.

White, C. M. N. and M. D. Bruce 1986: *The birds of Wallacea (Sulawesi, the Moluccas and Lesser Sunda Islands, Indonesia)*. London: B. O. U. Checklist no. 7.

Wit, H. C. D. de (ed.) 1959: *Rumphius memorial volume*. Baarn: Hollandia N. V.

Wit, H. C. D. de 1959: A checklist to Rumphius's Herbarium Amboinense. In *Rumphius memorial volume*. (ed.) de Wit, H. C. D.. Baarn: Hollandia N.V.

Wouden, F. A. E. van 1968: *Types of social structure in eastern Indonesia* (Koninklijk Instituut voor Taal-, Land- en Volkenkunde, Translation Series **11**). The Hague: Martinus Nijhoff.